ASTEROIDS

Kosmos

A series exploring our expanding knowledge of the cosmos through science and technology and investigating historical, contemporary and future developments as well as providing guidance for all those interested in astronomy.

Series Editor: Peter Morris

Asteroids

Clifford J. Cunningham

REAKTION BOOKS

Published by Reaktion Books Ltd
Unit 32, Waterside
44–48 Wharf Road
London N1 7UX, UK
www.reaktionbooks.co.uk

First published 2021
Copyright © Clifford J. Cunningham 2021

Printer and bound in India by Replika Press Pvt. Ltd

A catalogue record for this book is available from the British Library

ISBN 978 1 78914 358 4

CONTENTS

FOREWORD

When I first became interested in asteroids, sometime in the early 1970s, there was very little literature available on the subject. Even among the professional community, there was little interest in them. Research papers were few and far between, and mentions in popular publications were practically non-existent.

Fast forward fifty years and asteroids are hot! We now know that asteroids hit planets, sometimes catastrophically. Mankind owes the fact that it is the dominant life-form on this planet to an impact that wiped out the dinosaurs. Studies of the distribution of the orbits and sizes of asteroids give us clues about the dynamics of the early solar system and the effects of accretion versus collisions as a process for forming objects. We have discovered objects in locations that were unknown fifty years ago: some always closer to the Sun than Earth; asteroids that revolve about the Sun in the opposite direction to the major planets; objects that orbit entirely beyond the orbit of Neptune; and celestial bodies whose appearance lies somewhere between the star-like asteroids and the fuzziness of comets. Amateur astronomers in the 1980s and early 1990s were able to discover many new asteroids, but large professional observing programmes, intended primarily to find the objects that might hit us in the future, then swept up those that could be found by amateurs. Many of them switched to observing light curves to determine rotation periods and spin directions. Space probes have flown by or orbited more than a dozen asteroids, changing our view of them from mere points of light in a telescope into worlds in their own right. And more missions are planned.

I have been involved in tracking asteroids and comets professionally for thirty years, with another ten years before that as an amateur. Since 1990 I have seen the observational activity on asteroids increase from roughly 50,000 observations a year to more than 20 million. In 1990 we were tracking about 5,000 objects. We now track almost a million. That has been a challenging data management and computation problem.

Clifford Cunningham has written on asteroids for more than thirty years, with a special interest in the history of asteroids. This interest evolved into his PhD topic: a study of asteroids in England in the early nineteenth century. His more scholarly publications include a large number of volumes based on translations of correspondences between numerous early nineteenth-century astronomers on the topic of – you guessed it – asteroids.

Will this book tell you everything you need (or would want) to know about asteroids? No, that is not possible in a book this size. But it will hopefully whet your appetite to learn more. Enjoy.

Dr Gareth Williams, former Associate Director,
Minor Planet Center

ONE

THE STRANGER CERES

Asteroids are quirky. Just consider the title of this chapter. It comes from a chapter title of an 1877 book by William Oxley of Manchester in the north of England. His book was as far removed from science as it could possibly be. 'Are you a stranger?' asks someone known as the Recorder. 'I have lived on the planet Ceres,' is the response. 'I am not a departed Spirit, but am an inhabitant of that planet you call Ceres.'[1] The answer is so preposterous that one can easily miss the two dramatic assumptions it contains: first, the belief, widespread in the nineteenth century, that other celestial bodies (including asteroids) can be inhabited; and second, that Ceres is a planet.

The simple question 'what are asteroids?' has no simple answer. Are they at least partly metallic objects orbiting the Sun between Mars and Jupiter? Are they dead comets? Are they objects in the outer solar system that were subject to weathering by liquid water? Are they interstellar visitors? Are they companions of the planets, following or leading the planets in the same orbits? Are they pristine objects from the origin of the solar system? Asteroids are all these things and more, making them the most interesting and diverse diminutive objects in the solar system.

In essence asteroids are small objects, composed of silicates or metal, that can be found throughout the solar system. Comets are composed of silicates and ice. The major distinguishing factors between asteroids and comets are their source regions and the presence, or lack, of ice and frozen gases. Meteorites typically originate from either asteroids or comets, although a few come from the Moon or Mars owing to impacts on their surface by asteroids, meteoroids or comets. When they survive the descent through Earth's atmosphere, they are termed meteors. Asteroids, comets and meteors are now collectively known as small bodies.

The key word here is 'small'. They are also called 'minor planets' or, in French, *petites planetes* and, in German, *kleine Planeten*. The International Astronomical Union (IAU), which prefers 'minor planet' to 'asteroid', developed the unattractive designation 'dwarf planet' to encompass the largest of these objects. Thus Ceres, the first asteroid discovered, has in the twenty-first century assumed the guise of a dwarf, cut from the same cloth as Pluto. The very suitability of the IAU (whose membership comprises some 14,000 professional astronomers worldwide) to vote on and adjudicate such matters as designating Ceres and Pluto as dwarf planets is still hotly contested by many of those astronomers.[2]

It was Giuseppe Piazzi and his assistant Niccolò Cacciatore who discovered Ceres on New Year's Day 1801. Never before has a century been inaugurated with such an epoch-changing astronomical event. But on that evening, neither man knew they had just made a major discovery. Piazzi saw through his astronomical instrument atop the Royal Palace in Palermo a 'star' which his assistant duly recorded. It was part of their grand effort to create an improved star catalogue and for this a special kind of telescope – a vertical circle – was essential. Accuracy was also necessary, and for this purpose Piazzi had commissioned Jesse Ramsden in England to make a circle with a diameter of 1.5 m (5 ft). It was the best such instrument in the world.

Thus when on 2 January Piazzi saw this same star again, he was annoyed with Cacciatore, whom he thought had made a mistake in recording its position the night before. Cacciatore, a stickler for

Left: Giuseppe Piazzi, director of Palermo Observatory, discovered the first asteroid, Ceres, in 1801.

Right: Niccolò Cacciatore, Piazzi's assistant, was co-discoverer of Ceres in 1801. He became Director of Palermo Observatory in 1817.

precision, was just as annoyed with Piazzi for rebuking him for the error. On 3 January Piazzi repeated his mistake, accusing Cacciatore of error. 'I was rather piqued at this,' Cacciatore said later. Allowing for the politeness of the age, one may reasonably conclude he was enraged. It was not until 4 January, when Piazzi observed the star for the fourth time, that he realized Cacciatore was blameless: the star had clearly moved. Piazzi exclaimed with joy, 'We have discovered a planet!' In Piazzi's 1802 official account of the discovery, Cacciatore was airbrushed out of history. It was not until my own research was published in 2016 that Cacciatore was given his due recognition as the co-discoverer of Ceres.[3]

Piazzi suffered from overweening pride. It doesn't feel like a sin, but it is one of the seven deadly sins, and the early nineteenth-century English essayist William Hazlitt wrote that it 'scorns all alliance with natural frailty or indulgence'. In the case of Piazzi both of these played a role in the story of Ceres. Having observed Ceres for only 41 days, Piazzi's natural frailty intervened – he became seriously ill, which prevented him from observing Ceres any longer. He got no sympathy from Great Britain's Astronomer Royal Nevil Maskelyne, who excoriated him in most unphilosophical language. Nothing could hide Maskelyne's temper. Piazzi, he wrote, 'was so covetous as to keep this delicious morsel to himself for six weeks; when he was punished for his illiberality by a fit of sickness, by which means he lost track of it . . . What a deal this imprudent astronomer has to answer for!' Keeping in mind Piazzi was a monk, it is safe to assume that Maskelyne thought he had been punished by a higher power. As a religious man, Piazzi was surely aware that curiosity had for centuries been repudiated and allied with pride 'and therefore opposed to a seemly wonder at God's handiwork: this is St Augustine's complaint against the astronomers, who suffer from an excess of curiosity and a deficiency of awe'.[4]

On 19 September 1801 Johann Bode, director of the Berlin Observatory, wrote in despair to Wilhelm Olbers, an astronomer in Bremen, 'Why did [Piazzi] not let his assistant [Cacciatore] observe the planet in March, April and May? We could indicate its current position with more accuracy.' Why indeed? Did Piazzi not trust continued observations of Ceres to Cacciatore? This is unlikely, as Cacciatore was a careful observer. Rather, it was pride that was Piazzi's Achilles heel. Piazzi had come to regard Ceres as his personal possession, 'like something I own', as he himself wrote. By indulging in this wholly misplaced paternal feeling, and falsely confident in his ability to determine its orbit, Piazzi lost Ceres.

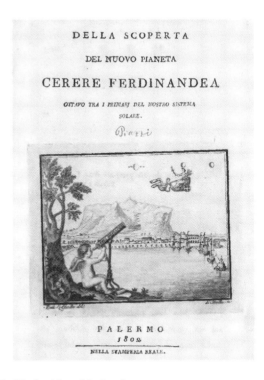

The 1802 report by Piazzi on his discovery of Ceres. The graphic depicts a cherub looking through a telescope at the goddess Ceres riding her chariot in the sky. In the background is Mount Pellegrino, which dominates the skyline of Palermo, the city where the asteroid Ceres was discovered.

Left: Nevil Maskelyne, Great Britain's Astronomer Royal from 1765 to 1811.

Right: Johann Bode, director of Berlin Observatory. In 1802 he wrote the first book about asteroids.

Far from being simply the first asteroid, Ceres proved to be a catalyst for a host of volatile issues, both scientific and personal. Its discovery and early study strained or destroyed friendships, both sullied and launched professional careers, upended the neat division between planets and comets, and fulfilled a tantalizing promise that was first propounded in 1595. That is when Johannes Kepler made the tentative suggestion that there was an unseen planet between Mars and Jupiter. He made the suggestion purely as a hypothesis to fit a regular order he perceived in the distances of one planet from another. This order was none other than the 'harmony of the spheres', a notion as beguiling as it was misleading. But this search for harmonious proportions in the solar system (in Kepler's case a series of nested geometric solids that defined the planetary orbits) set the stage for others to ponder further about the large gap that separates the orbits of Mars and Jupiter.

In 1681, nearly a century after Kepler's suggestion, the great Swiss mathematician Jacob Bernoulli made a bold claim. Not only was there a planet between Mars and Jupiter, but Bernoulli could and did print its orbital elements. He published his findings in the Dutch language, which had a limited readership. These days a researcher would publish a work in English to attain a wide readership; in 1682 Bernoulli chose Latin for a new edition of his work, as that language was Europe's intellectual lingua franca. Even so, it appears no one took any notice, certainly not enough to launch a search. The matter rested for a century.[5]

Then a young Hungarian-born astronomer, Franz Xaver von Zach, wrote down the orbital elements of the 'missing planet'. He dared not publish these figures, but instead deposited sealed envelopes with three people. That was in 1785, and it was (in his own words) a 'dream' he could not wake from: the dream of finding another planet. Fifteen years later he finally made good on his intentions by founding the United Astronomical Society for the purpose of searching the skies for the planet. The Society is known to posterity as the Celestial Police (a popular name given to it by Zach in 1801), but it never engaged in serious work, as Piazzi discovered Ceres quite by accident less than four months after its formation.

Franz Xaver von Zach, director of Seeberg Observatory, was the first to recover Ceres in December 1801 after it had been lost from view earlier in the year by Piazzi.

Zach, however, was to savour a very special moment when he was the first to see Ceres after it had been lost by Piazzi. His sighting was made on 7 December 1801, but it was not until 1 January 1802 that he was able to confirm the rediscovery. He also happened to be the editor of the world's only monthly journal of astronomy, and it was in the pages of the German-language *Monthly Correspondence* that most of the observational papers about the first asteroids were published.

In his search to rediscover Ceres, Zach was guided by the predictions of the German mathematician Carl Gauss as to where Ceres could be found after an absence of ten months. Why did the learned astronomers of Europe have to rely on a 24-year-old, and until then virtually unknown, mathematician to end their suffering? For suffering it was, as letter after letter between those astronomers in 1801 bewailed their wasted efforts as they searched in vain for the lost Ceres.

To answer the question, we must consider the role of mathematics at the time. Many prominent researchers, including Benjamin Franklin and the Frenchmen Denis Diderot and the Comte de Buffon, dismissed the application of mathematics to physics. Their understanding of nature did not rely on the abstract forms of mathematical equations. But some of their colleagues in France adopted the opposite view, including the Marquis de Condorcet and Jean d'Alembert. From 1751 to 1772 a multi-volume work titled *Encyclopédie* was published in Paris. According to Denis Diderot, it was meant 'to change the way people think'. (Diderot and d'Alembert were co-editors in the early years of publication.) In the 'Preliminary discourse' to the *Encyclopédie*, d'Alembert wrote that mathematical knowledge was equally at home in our examination of terrestrial and astronomical bodies.

> All the properties we observe in these bodies have relationships among themselves . . . The knowledge or the discovery of these relationships is almost always the only object that we are permitted to attain, and consequently the only one that we ought to purpose for ourselves.[6]

Owing in great part to his efforts, people did change the way they thought about the role of mathematics in the study of nature. By the time Ceres was discovered in 1801, a mathematical approach known as celestial mechanics had been accepted as the only way to understand what many regarded as the clockwork mechanism of the solar system. When Gauss became aware, in mid-1801, of Ceres being lost, he resolved at once to find a mathematical solution. All he had to work with were the numbers given by Piazzi that recorded the time of observation and the coordinates in the sky where Ceres was seen.

Numbers assumed a new importance in the Enlightenment. In the words of the professor of German Martin Dyck, 'In the domain of things visual or geometrically visualized, numbers pinpoint the elusive contours, configurations, or proportions of beauty and order rendered physically discernible.'[7] Fortunately for astronomy, the mathematical genius of Gauss positioned him as the only person fully capable of grasping the beauty and order of the numbers that led to the place where Ceres could become physically discernible. Others laboured to calculate the orbit of Ceres and its place in the autumn sky of 1801. Astronomers looked there and found nothing. Gauss developed a new mathematical approach that yielded a different position in the sky, and it was there Zach found it in December 1801, the recovery being confirmed by Olbers in January 1802. Piazzi himself was most grateful to Zach for finding his cherished Ceres: 'I can never express what feelings of gratitude and attachment all of your efforts have aroused in me. I find in you such a brother and friend I could not possibly have hoped to find any better.'[8] As John Rees floridly stated in a popular lecture before the New York Academy of Sciences in 1897, 'The "celestial police" had at last arrested the fugitive and had taken her picture for the rogues' gallery; she could never thereafter get away again.'[9]

Onofrio Piazzi, a distant relative of Giuseppe with shared ancestors dating back to the sixteenth century, wrote in verse on the discovery of Ceres. Here are a few lines from the poem, published in Milan in 1829, the year after Onofrio's death.

Let your memorable name be that of
Ferdinandean Ceres; and the august name
Ferdinand I was seen to shine
in the new planet itself.[10]

Piazzi had named his discovery both for the patron goddess of Sicily and for his own patron, King Ferdinand I of the Two Sicilies. While only a few people called the object Ceres Ferdinandea, the name does reflect the important point that the existence of many observatories of that age was due to the largesse, and foresight, of royalty.

This was not the case, however, for the observatory of Olbers. Indeed, he didn't actually have an observatory at all: his telescopes were placed in a small room at the top of his house. By day he practised medicine downstairs and at night he practised astronomy upstairs. While looking at Ceres in March 1802, Olbers made an extraordinary discovery: a second object orbiting between Mars and Jupiter. Olbers named it Pallas. Clearly, Ceres could not be the missing planet if there was another one! The excitement at the time can be judged from the opinion of Zach, who wrote of Pallas that it was 'among the most important discoveries ever made . . . an extraordinary stranger in the solar system'.[11]

Bode's Law

The expectation that just one missing planet would be found rested in part on what became known as Bode's Law, named after Johann Bode. Without getting into the mathematics, Bode's Law was able to predict the distance of each planet from the Sun. It was first formulated before the discovery of Uranus, and when that planet was found by Herschel in 1781 its distance from the Sun fitted nicely into the progression expected. But it also predicted a planet between Mars and Jupiter, the gap that had already been noted by Kepler and

Bernoulli. The whole concept was also bound up with the mystical notion, present since the times of the ancient Greeks, that there was a 'music of the spheres' that expressed harmony in the cosmos.

While Bode always couched it in terms of a hypothesis, others seized upon it with such fervour that it came to be regarded by many as a law, which it certainly was not. Newton's Law of Gravity, for example, is a true law of nature, one that can be verified experimentally. Bode's relationship, as it should be called, was just empirical. Even Johann Wurm, who in 1787 created an equation to express Bode's Law in mathematical terms, raised doubts about it. In German he termed it an 'astronomische Schwärmerei'. Schwärmerei, just like love, is something you believe in just because you want to believe in it, and therefore you'll find countless plausible 'excuses' and explanations to demonstrate that it exists, even though any outside person, without the emotional attachment, might simply call it 'fantasy' (whether that is true or not). If subsequent researchers had heeded Wurm's assessment of the 'astronomical fantasy' it would have saved a lot of controversy over the next two centuries.

The danger inherent in relying on hypotheses (such as Bode's Law and the belief shared by several writers of the age that asteroids were inhabited) was neatly summed up by Professor James Dean of the University of Vermont on 24 April 1810. It is worth quoting as a warning that is just as relevant today as it was two centuries ago.

Let us not permit our curiosity to be gratified by building flimsy ephemeral hypotheses, whose principles, like the figure of the beauteous snowflake, vanish under the gazer's breath. But, comparing fact with fact, as long as facts are to be found, let us subject every suggestion of the imagination to the strictest rules of mathematical reasoning.[12]

The philosopher Philo, who lived in Egypt and died around the year 50 BCE, was quite prescient in laying out the mindset that would enable Bode's Law to develop 1,700 years later. He wrote of a 'sober intoxication' that seizes the mind 'whirled round with the dances of planets and fixed stars, in accordance with the laws of perfect music', and so at last 'descrying in [the intelligible] world sights of surpassing loveliness, even the patterns and originals of the things of sense which it saw there'.[13] The planets are indeed sights of surpassing loveliness, and taking the step of discerning a pattern that governed their distance from the Sun was a natural expression of the desire to explain and codify their existence. How asteroids fit into this pattern was an unexpected problem astronomers of the early nineteenth century were forced to grapple with. With little else to go on, most adopted Bode's Law either cautiously or whole-heartedly.[14] As Nobel laureate in physics Steven Weinberg told me in 2018, 'I don't think one loses much by proposing a hypothesis that does not conform to nature – Nature gets the deciding vote in telling us whether we are right or wrong.'

Ceres and Pallas: Planets or Asteroids?

Ceres and Pallas were readily accepted as new planets, but the reality is more complex than this. Piazzi himself equivocated throughout most of 1801 as to whether he had discovered a planet or a comet. There was precedent for this, as William Herschel believed he had discovered a comet in 1781, when in reality he had found the planet later named Uranus. Herschel was led to this conclusion because he measured the object as nearly doubling in size over a period of one month. In reality Uranus only grows by about 10 per cent over the course of a year, so this spurious set of measurements led Herschel astray.

Since Piazzi's object remained unseen by anyone but himself and Cacciatore as 1801 progressed, the great astronomer and

William Herschel made the first scientific study of asteroids, which was published in 1802.

mathematician Pierre-Simon Laplace in France publicly raised doubts about its very existence. By contrast, Bode knew as soon as he received the discovery letter from Piazzi on 20 March 1801 that a new planet had been found. To attribute this to a leap of faith would be incorrect. His conclusion is more properly encompassed by the French verb *sentir*, which connotes an immediate and spontaneous understanding or knowing, an affective intuition combined with an act of intellection. But the capacity for *sentir* was absent in 1802: when it became obvious Pallas had a very high inclination, Bode was convinced it was a comet. Why? Because he wanted to deny planetary status to Pallas as it might cause his hypothesis to be discredited. Thus when his fellow German astronomer Reissig, living in Kassel, reported seeing a new object in February 1803, Bode never entertained the notion it might be yet another planet. Reissig wrote to Bode, 'The star or comet appeared without appreciable nebulosity and slightly enlarged when viewed at 400×.' Reissig observed it as it moved, seeing it on four nights. In his annual publication the *Berlin Astronomical Yearbook*, Bode wrote that the object was probably a distant comet in retrograde motion. In the late twentieth century, Brian Marsden suggested it was quite likely an asteroid that passed close to Earth. Such an object would have been inconceivable to astronomers in 1803, so this potentially amazing discovery was quickly forgotten.

Why did the opinions of well-respected astronomers not persuade everyone that only comets had been found in these early years? Simply put, astronomy did not have any organized body that could pass judgement. While Laplace and Piazzi were exemplars of astronomy, they were not its guardians, and their opinions failed to persuade Zach and Olbers. If they too had shared Laplace's view that Piazzi had not really discovered a moving object, they would never had tried searching for it, but their conviction that it was a planet made the search all the more urgent. Likewise, Bode's conviction that Pallas was a comet gained no traction, despite his eminent status. The practice of intellectual labour continued despite all the recriminations and doubts. Ceres and Pallas were real, and they were planets. Or were they?

How the Term 'Asteroid' was Born: A Timeline in 1802

Herschel was of the opinion that Ceres and Pallas were not planets, but what to call them? Herschel faced the same dilemma that confronted Portia, a character in Sarah Fielding and Jane Collier's novel *The Cry* (1754). She believed 'the change of words, and the adapting them to new purposes, doth but make language follow the fate of all nature'.[15] But for Herschel the word 'planet' could not be repurposed to encompass Ceres and Pallas. He adopted the same approach as Portia: new objects require new words. In the novel she invented such words as 'dextra' and 'turba' when existing vocabulary failed to convey the truth-describing essence of what she needed. Herschel likewise needed a word to describe the new truth about the solar system as revealed by Ceres and Pallas, but he was no linguist.

On 25 April 1802 Herschel wrote to his friend William Watson asking him to create a term that would tell the world Ceres and Pallas were not planets. Two days later Watson replied with a list of names, including Planetel (Watson's personal favourite), Planetkin and Planeret. Herschel quickly rejected these, turning to another friend, Charles Burney Sr. On 3 May Herschel paid a personal visit to Burney to appeal for an appropriate term. Later that day Burney sent a letter to his son, a noted expert in ancient Greek, suggesting something containing 'aster' might work. The very next day, Charles Burney Jr created the word 'asteroid', and sent it to Herschel. 'Asteroid' is derived from Greek. It simply means star-like; the suffix -oid actually means 'having the nature of', but the word Burney Sr specifically directed his son to create was meant to denote 'a small star'.

On 6 May Herschel's paper on Ceres and Pallas, the first scientific paper ever written about the asteroids, was read in London at a meeting of the Royal Society. It incorporated the term 'asteroid', but he was not fully satisfied with it so he appealed to Sir Joseph Banks, President of the Royal Society, to come up with a better appellation. On 7 June Banks sent Herschel a list of potential 'Names for the new planet', writing:

> I applied to Mr S[tephen] Weston as I always do in these occasions to stand God Father to your new species of moving stars and [he] has sent me a card which I enclose. I really think Aorate a good name and much better than any that has been hitherto suggested and the more so as it is not probable that any of this new kind of wanderers are visible to the naked eye.[16]

On the same day that he sent the list to Herschel, Banks wrote to Baron von Zach that Herschel had decided to use the term 'Aorate'. It was a classic case of jumping the gun, as Herschel had made no such decision. Banks's letter was published by Zach in his journal, the *Monthly Correspondence*:

> Dr Herschel still persists in his opinion in view of the small size of these two new planets, and continues to maintain that these must be strictly differentiated and classified especially from planets and comets, except when these questionable comets are found in a quiescent state. I believe that he wishes to name them Aorates, because they are not visible with the naked eye. We see no difficulty in the requirement that the light from such small bodies should reach us.[17]

On 10 June Herschel rejected Stephen Weston's word 'Aorate' in this letter to Banks, but it was too late to stop Zach publishing Banks's letter.

The names you have done me the favour to send I have carefully examined, and beg leave to give you my remarks on them. The title of them, 'Names for the new Planet,' shews immediately that none of them can possibly be used for the new species of bodies which we have to christen: for they are not planets.[18]

Even given Herschel's authority as the only discoverer of a planet in recorded history, no one except Olbers approved of his decision to demarcate Ceres and Pallas from the much larger planets. Virtually every book and paper until the mid-nineteenth century called the new discoveries 'planets'. Herschel's son John awkwardly referred to them as 'ultra-zodiacal planets'.

Piazzi believed them to be planets but reluctantly suggested in a letter to Herschel in 1802 the word 'planetoid', which garnered some support: in his last book, published in 1928, the American geologist Thomas Chrowder Chamberlin wrote, 'the old name "asteroids" is such a monstrosity that we cannot use it in the face of so appropriate a term as "planetoids".'[19] A publication of 1901 has the silliest description: 'A broad stripe of the heavens on each side of the ecliptic is tattooed with these little "pocket planets".'[20] If one is looking for an 'official name' it must be minor planets, although the IAU now lumps asteroids, comets and meteoroids into the catch-all phrase 'small solar-system bodies'.

Of the astronomers who observed asteroids in the early nineteenth century, few made an attempt to determine anything about their physical nature. While others were content to measure their positions and compute ever better orbits, a brave handful pushed the limits of technology to find out what they really were. In 1802 William Herschel published the first scientific paper on asteroids based on observations with his 20-foot (6 m) telescope. He measured their diameters, and was followed in this by Johann Schroeter and Olbers. In a report dated 2 November 1804, shortly

Herschel's telescope, with a 20-foot-long (6 m) tube, was used in 1802 to study Ceres and Pallas from Slough (near Windsor), England.

after the discovery of the third asteroid, Juno, Olbers gave his estimate of the relative sizes of the first three asteroids, without actually assigning them diameters.

> Juno is most likely the smallest of the three known asteroids. If it were as big as Ceres, it would have had, given its position to the Earth and the sun in September, five times the luminosity of Ceres during the same period. But its luminosity was only slightly greater than the lower-lying Ceres. I conclude from this that it is only about half the size of Ceres in diameter. I have estimated the diameter of Pallas as two-thirds the diameter of Ceres using similar photometric comparisons.[21]

But on 1 January 1805, he reported a larger relative diameter of three-quarters for Pallas: 'If it is assumed that the albedo [proportion of reflected sunlight] of these three new planets are the same or that the surface of all three reflect the sunlight equally well, then the relationship of their diameters will be very close to the following: Dia. of Ceres = 1.00; Pallas = 0.74; Juno = 0.43.'[22]

Olbers managed to discover the truth that Ceres was the largest, followed by Pallas and Juno, but his proportions were mistaken. Given 945 km (587 mi.) for Ceres as the correct value, he was saying Pallas was 700 km (435 mi.) and Juno was 400 km (248 mi.). The true diameters are 511 and 258 km (317 and 160 mi.), respectively. As a first approximation, the assumption of a consistent albedo is a logical one, but it led Olbers astray. In reality only 9 per cent of the light hitting Ceres from the Sun is reflected; for Pallas it is 10 per cent but for Juno it is 24 per cent. To round out the list of the so-called Big Four, Vesta (discovered in 1807) is 525 km (326 mi.) with a very high albedo of 42 per cent.

The term 'Big Four' has often been used to describe Ceres, Pallas, Juno and Vesta collectively, but it is a misnomer, as seven other main belt asteroids are bigger than Juno. The 'Big Four' are really just the 'First Four'. Schroeter was misled into thinking the asteroids had large atmospheres, so his estimates of their sizes were far too big (consider the astounding 3,258 km (2,024 mi.) suggested for Pallas), whereas Herschel measured them to be much smaller than they really are (for instance, 236 km (147 mi.) for Pallas). Reasonably accurate determinations of the sizes of the largest asteroids, measured by instrumentation more advanced than anything available in the nineteenth century, only became available in the 1950s and '60s. Ceres is the largest main belt asteroid to assume a spherical shape, like the major planets, so whether one regards Ceres as a planet, dwarf planet or asteroid is a matter of interpretation. Chapter Two explains what the main belt is, and where the other spherical asteroids are located.

TWO

WHERE ARE THE ASTEROIDS?

Until the second half of the nineteenth century, fewer than fifty people in the world had actually observed an asteroid through a telescope. Just getting an accurate position of one of the known asteroids so that one could find it was very difficult. In England the *Nautical Almanac* was an annual publication that was relied on by astronomers for a host of data on astronomical objects, but it was sadly lacking when it came to asteroids. This even prompted a letter to *The Times* newspaper of London on 9 April 1830, in which someone complained that 'planets discovered 30 years ago are still not mentioned' in the *Almanac*. The writer said he had to rely on the German counterpart, the *Berlin Ephemeris*, for asteroid data. (Yearly ephemerides for the asteroids were published by Berlin until 1891, when most were allotted just one line giving the time and place of opposition, and daily variation in position; important asteroids still received daily positions for six weeks of their best visibility.)

The discovery of a fifth asteroid, after a gap of 38 years since the discovery of Vesta, was hailed by the noted English astronomer James South. In a letter to *The Times* published on 29 December 1845, he wrote that Karl Ludwig Hencke 'has rendered a service to astronomy of indeed high moment'. But the new object, dubbed Astraea, was not greeted with universal enthusiasm. In 1847 a writer stated: 'The interest of the discovery of an obscure asteroid sinks into insignificance when compared' with the 1846 discovery of the planet Neptune. Lack of appreciation did not dissuade Hencke, who went on to discover the sixth asteroid, Hebe, in 1847. Hencke's success was hard fought: he had begun searching for more asteroids in 1831.

By the next decade asteroids had become the hot topic in astronomy. In 1853 the *North British Review* claimed there was no branch of astronomy where the progress of discovery had been more rapid: nineteen new 'planets' had been found between 1845 and 1852. The Russian writer Aleksei Savich thought highly of these discoveries. In a book of 1855, he wrote:

> The discovery of a new planet or comet has a similar significance in astronomy as does in botany the description of new types of grasses. Without doubt these discoveries are fully deserving of gratitude, in part because of their difficulty, and in part because they increase the material for knowledge. Their work is not easy, and far be it from us to consider their activities unimportant. Therefore no one will begrudge praise being given to those workhorses of sciences who, sacrificing their peace, spend whole nights looking for these barely noticeable bodies.[1]

One of the greatest of these workhorses was the German astronomer Robert Luther, who discovered 24 asteroids between 1852 and 1890. In honour of his first discovery, Thetis in 1852, he was fêted at a banquet in Dusseldorf's Knight's Hall. This changed his life forever. In 1876, after he had worked in Dusseldorf for a quarter century, Luther made a remarkable admission.

> As I do not like at all any noisy festivities, such as those presented to me here after the discovery of Thetis, I have kept strict silence about my 25-year-long presence here. A friendly letter was much more appreciated than all the festivities which are even harmful for health and the capacity to work.[2]

The asteroid discoverers of this era did not work entirely in isolation: many of them were acquainted with one another. Before his move to Dusseldorf, Luther worked in Berlin from 1843 to 1851. Johann Galle, director of the Berlin Observatory, remembered Luther after his death in 1900.

> His Berlin University days coincided with the beginning of the new discoveries in the field of the asteroids, namely the discovery of Astraea by Hencke which at that time was accompanied by such a sensation and excitement as is hardly remembered in connection with the numerous later discoveries in this field. In Berlin, Luther had simultaneous studies and activities with Franz Bruennow and Heinrich d'Arrest at the Berlin Observatory.

Bruennow wrote numerous papers on the orbits of asteroids, and in 1862 d'Arrest discovered the 76th asteroid, Freia. In Berlin Luther also met Eduard Vogel, who became an assistant to John Russell Hind in 1851. In December 1855 Luther was decorated with the order of the Red Eagle Class IV by King Friedrich Wilhelm IV and, in 1863, with the order of the Red Eagle by King Wilhelm I. He was also awarded an honorary doctorate and the Lalande Prize

Asteroid medal of the French Imperial Institute, France, in honour of the astronomers John Russell Hind, Hermann Goldschmidt and Robert Luther; illustration from the *Illustrated London News*, vol. LV (9 October 1869).

PLANETARY MEDAL OF THE FRENCH IMPERIAL INSTITUTE.—SEE PAGE 368.

of Astronomy seven times for his discoveries. Despite his retiring nature, Luther reached an eminence due entirely to asteroid discoveries that was never surpassed. But he was not alone in racking up the asteroid count: James Craig Watson in Ann Arbor, Michigan, found 22; Christian Heinrich Friedrich Peters at the Litchfield Observatory in New York found 48; Johann Palisa at the observatory in Pula (then part of the Austrian Empire) found 122; and Auguste Charlois in Nice discovered 99. Max Wolf in Heidelberg found more than two hundred, mostly by photography. In 1869 a medal was struck by the French Academy to commemorate the discovery of the hundredth asteroid. It featured portraits of Hind (England), Goldschmidt (France) and Luther (Germany).

The English and French, though rivals in the discovery of Neptune, joined forces with the United States to keep an eye on asteroids. George Airy, at the Greenwich Observatory, felt that it was too great a burden for his observatory alone. In 1863 he made an agreement to share the task with the U.S. Naval Observatory, thus establishing one of the earliest cooperative observing programmes in astronomy.

By the 1860s so many asteroids had been found that mapmakers gave up trying to show their individual orbits, merely placing a few dots between Mars and Jupiter and adding a table to list them all. An author in 1860 went so far as to denigrate the intellectual ability of those who discovered asteroids. 'The discovery of new asteroids is like the discovery of nuggets by the gold-digger. The star-finder, in such a case, claims none of the prophetic genius of Leverrier and Adams.' He was referring here to the French and English astronomer-mathematicians whose work led to the discovery of the planet Neptune. By the turn of the century, opinion had soured still further. 'It is a great task to take care of so large a family (nearly 475), and astronomers now look with disfavour on their further discovery,' an annual American publication reported in 1901. They earned a dubious reputation as 'vermin of the skies',

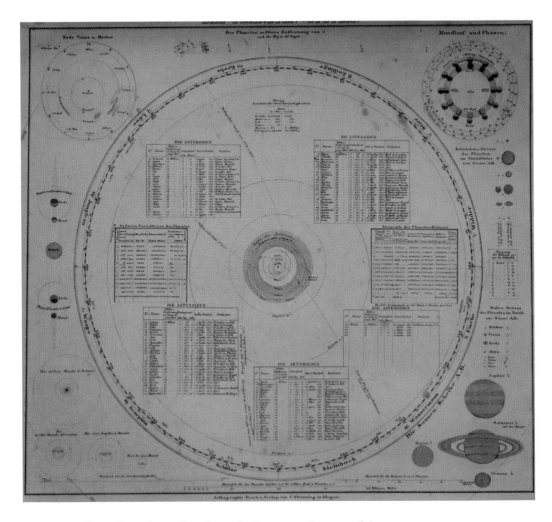

a term attributed to Edmund Weiss, who became director of the Vienna Observatory in 1878. In 1900 the astronomer Viktor Knorre explained the situation as the twentieth century began:

I can add from my own experience that the sparkling effect which the results in the field of discovering small planets initially had on the anxious public interest – of which I have repeatedly seen notices in the scientific journals – soon waned.

Map of the solar system published by the German cartographer Carl Flemming in 1868, containing data on 106 asteroids in five tables. The zone where asteroids orbit is indicated in grey.

Detail of the map published by Carl Flemming, showing one of the five tables of asteroids. This was the last map to give such detailed information, as the number of asteroid discoveries quickly made this approach impractical.

DIE ASTEROIDEN.

N°	Namen	Mittlere Entfernung v.d. Sonne in geogr. Meil.	Umlaufszeit Jahr	Tag	Stdn	Zeit d. Entdeck.	Entdecker
53	Calypso	34 Million.	4	79	1	4 April 1858	Dr. Luther in Bilk
23	Thalia	54 "	4	92	4	15 Decbr. 1852	Hind in London
37	Fides	55 "	4	107	5	5 Octbr. 1855	Dr. Luther in Bilk
15	Eunomina	55 "	4	108	3	29 Juli 1851	de Gasparis, Neapel
50	Virginia	55 "	4	113	13	4 Octbr. 1857	Ferguson Washington
26	Proserpina	55 "	4	119	20	5 Mai 1853	Dr. Luther in Bilk
3	Juno	55 "	4	132	15	1 Septbr. 1804	Dr. Harding, Lilienthal
34	Circe	56 "	4	148	16	6 April 1855	Chacornac, Paris
54	Alexandra	56 "	4	166	1	10 Septbr. 1858	Goldschmidt, Paris
45	Eugenia	56 "	4	156	9	27 Juni 1857	ders.
38	Leda	57 "	4	195	1	12 Jan. 1856	Chacornac, Paris
36	Atalante	57 "	4	203	6	5 Octbr. 1855	Goldschmidt, Paris
55	Pandora	57 "	4	213	2	10 Septbr. 1858	Searle in Albany
1	Ceres	57 "	4	218	13	1 Jan. 1801	Piazzi in Palermo
2	Pallas	57 "	4	221	19	28 März 1802	Dr. Olbers in Bremen
39	Lätitia	57 "	4	223	13	8 Febr. 1856	Chacornac, Paris
28	Bellona	57 "	4	227	8	1 März 1854	Dr. Luther in Bilk
33	Polyhymnia	59 "	4	309	14	28 Octbr. 1854	Chacornac in Paris
47	Aglaja	60 "	4	326	19	15 Septbr. 1857	Dr. Luther in Bilk
22	Calliope	60 "	4	351	8	16 Novbr. 1852	Hind in London
16	Psyche	61 "	5	1	20	17 März 1852	de Gasparis Neapel
35	Leukothea	62 "	5	46	9	19 April 1855	Dr. Luther in Bilk
49	Pales	64 "	5	154	14	19 Septbr. 1857	Goldschmidt Paris
52	Europa	64 "	5	168	14	4 Febr. 1858	ders.
48	Doris	64 "	5	173	13	19 Septbr. 1857	ders.
10	Hygiea	65 "	5	214	18	12 April 1849	de Gasparis Neapel
24	Themis	65 "	5	215	7	5 April 1853	ders.
31	Euphrosine	65 "	5	221	65	2 Septbr. 1854	Ferguson, Washingt.

When finally the discoveries reached a speed at which the experts started to lose courage in view of the increasing number of observations and calculations which are necessary for recording these celestial bodies, the initial thrill of novelty made room for a general indifference; one could even notice that this superfluous activity was met with a pitying smile.[3]

By 1912, Joel Metcalfe (who discovered 41 asteroids from 1904 to 1914) wrote, 'Formerly the discovery of a new member of the solar system was applauded as a contribution to knowledge. Lately it has been considered almost a crime.' Just the bookkeeping, much less the orbital calculations, proved burdensome as more and more were found. From the beginning of the asteroid saga in 1801, asteroids were assigned both names and symbols, the latter being useful as a shorthand in texts and tables. In 1577 Claude Mignault wrote a *Treatise on Symbols*, in which he stated that symbols should 'contain in dignified and brief form a great deal of meaning', but should not be 'so veiled and obscure' that one needs a seasoned diver to plumb their depths. Symbols for the first asteroids adhered to these principles. For example, Ceres was given the symbol of a sickle, representing the fact that in Greek mythology Ceres was the goddess of agriculture and the sickle was used to cut wheat. Vesta, goddess of the hearth, was represented by an altar topped by a sacred fire (a symbol devised by Gauss himself). As time went on, however, the most outlandish symbols, essentially devoid of meaning associated with the names, were created. Only the first twelve were assigned symbols by astronomers, after which each asteroid was known by its name and number. More recently, newly discovered asteroids are assigned a provisional designation. For example, 2004 MN_4 was found in 2004 in the last half of June (M). The subscript number is appended to indicate the number of times that the letters from A to Z have cycled through (the letter I is omitted in both letter sequences, leaving 25 letters). The suffix '4' indicates 4 completed cycles (4 cycles × 25 letters = 100), while N is the thirteenth position in the current cycle. Once an orbit becomes established, an asteroid receives a permanent number, and then becomes a candidate for a name. As of August 2020 some 990,091 asteroids had been discovered; 546,077 of these are numbered, but only 22,129 have officially been assigned names.

Piazzi was the first to both find and lose an asteroid (Ceres), but he was certainly not the last. Consider this story, colourfully told by William Fitzgerald in 1896.

> Of new minor planets, you are told proudly, thirty-three were discovered in 1893; and several lost planets – poor things! – were re-discovered by our celestial police, and conducted safely into the observatories – or, at any rate, records of them. But let

me illustrate this pretty game of hide and seek. Minor planet Sappho, who had maintained a decent position in heavenly society for years, suddenly disappeared. Her description was known to proper authorities, and the Press took up the case. At last Dr Isaac Roberts, F.R.S., got a clue, and set a photographic snare for Sappho. She fell into it.[4]

This amusing tale actually represents a watershed time for asteroid discoveries. No longer was the lone observer peering through a telescope the only – or even the best – way of finding asteroids. It was a nineteenth-century example of technology overtaking human achievement. It was Max Wolf who pioneered the photography of asteroids on 28 November 1891. With a 6-inch (15 cm) lens, his photo-telescope (as it was termed in the 1890s) could image 70 square degrees, and with exposure times of two hours he could detect asteroids fainter than the unaided eye could see by simply peering through a telescope. This revolution in the discovery of asteroids was marked in a lecture delivered in 1895 by Rev. Edmund Ledger, professor of astronomy at Gresham College in London. He noted that of late the rate of their discovery had increased rapidly: in the past four years, 68 of the 71 new asteroid discoveries were attributed to photography. The method Ledger showed his audience remained essentially unchanged throughout the twentieth century. He exhibited two plates of the same patch of sky, taken one night apart: the stars remained in their places, but the asteroid betrayed its existence by moving its position. From there it was a small step to create a device known as the blink comparator (invented in 1904), allowing one to flip rapidly from one image to another. Anything moving would be fairly easy to spot, and it was with such a device that Pluto was discovered by Clyde Tombaugh in 1930. From then until the present, telescopes using photographic plates or CCDs, and more recently spacecraft, became virtually the only method of discovering asteroids.

Today we find asteroids in many areas of the solar system, and each type has a name to distinguish it. Where the asteroids are located can tell us much about the early evolution of the solar system, so their study is of the greatest importance. Asteroids, while more numerous than ever, are no longer the vermin of the skies. The astronomers of the nineteenth century found nothing less than a celestial cluster within the solar system, a discovery that ranks among the greatest in astronomy. But only in the last few decades has their importance been fully recognized. As explained by Don Davis of the Planetary Science Institute:

> Asteroid science, along with the rest of planetary science, was pretty much an astronomical backwater in the decades prior to 1960. Since then, we have gone from knowing of a few thousand asteroids to nearly a million. Back in 1960 we saw asteroids merely as points of light in telescopes, and now they are being visited by spacecraft. The field of asteroid science has enjoyed a phenomenal expansion in recent decades.

Until 1873 the existence of asteroids beyond the range limited by the orbits of Mars and Jupiter – the main belt – was unknown. This state of affairs was neatly expressed by James Challis of Cambridge Observatory, who wrote in 1870 to the Rev. B. S. Clarke: 'I will not disguise from you my opinion that the theory of asteroids has little foundation to rest upon. The small planets revolving about the Sun between Mars and Jupiter being excepted, we have no positive evidence that any small *bodies* circulate about the Sun.' Now we estimate that millions of them exist into the farthest reaches of the solar system.

We have already encountered Bode's hypothesis about planetary distances. In the 1860s another hypothesis arose to explain asteroid distances from the Sun; the development of this hypothesis had a very different fate. The American astronomer Daniel Kirkwood initially studied the orbits of 87 asteroids, noting that they were not distributed evenly between Mars and Jupiter. The first statement he made about gaps in the belt was before a meeting of the American Association for the Advancement of Science in Buffalo in 1866. Within two years he published his findings based on 97 asteroid orbits, and posited that places between Mars and Jupiter that were without asteroids were due to 'the action of Jupiter'. By 1883 Kirkwood had noted the discovery of 120 more minor planets 'and, as was then predicted, the chasms in the zones have been rendered the more obvious'. We now refer to these as Kirkwood gaps, confirming his belief that the gravitational influence of the giant planet Jupiter is responsible for them.

The gaps are a result of the phenomenon of resonance, a term taken from acoustics where it was observed that a note struck on a piano can set a nearby violin string vibrating. Considering Jupiter as the celestial piano, and the asteroids as violin strings, a series of resonances were created in the early solar system. For example, in the orbit where an asteroid would go around the Sun three times in the same interval in which Jupiter would revolve just once, we find a gap in the asteroid belt. Some resonances are stronger than others, and three of them are not gaps at all but rather areas where asteroids concentrate. Here we find the Hilda group (at the 3:2 resonance, meaning their orbital period is two-thirds that of Jupiter), the Thule group (at 4:3) and the Trojans (at 1:1).

Some asteroids are highly inclined, while most are not. Inclined to what? As Dan Seeson neatly explained in 1955, assume the Sun to be floating in the centre of a calm pool, half of it under water; then, assume that the Earth, also half-submerged, floats around the Sun on an oval course; then the surface of the water could be said to represent the 'plane of the ecliptic'.[5] This has no reference to the position of the Earth's equator, nor to the imaginary projection of it upon the visible heavens (the celestial equator); it refers only to the path of the Earth around the Sun. Pallas follows an orbit inclined

35 degrees to that plane, a fact that caused considerable consternation in those who first studied it. Some believed that no planet could be so inclined, so it must be a comet. Pallas held the record for more than a century, until 1920, when 944 Hidalgo was found to have an inclination of 42.5 degrees. Some one hundred asteroids are now known to have inclinations higher than 90 degrees, which means they exhibit retrograde motion.

Near-Earth Asteroids

In 1873 the 132nd asteroid was found. Even though it was lost after only three weeks of observations (it was recovered in 1922), Aethra made history: it spends a small portion of its time closer to the Sun than Mars, the first time an asteroid was seen to deviate from the main belt. It was the discovery of 433 Eros in 1898 that expanded our view of asteroids in particular and the solar system in general. Aethra crosses the orbit of Mars, but Eros is more remarkable: it spends most of its time closer to the Sun than Mars. It was visited by NASA's NEAR spacecraft in 2001, the first space probe to land on an asteroid. Where else might asteroids be found? In a book of 1898, the poet James Slimmon even raised the possibility of a 'lunar asteroid'. We now know of several populations of asteroids that cross the orbit of Earth. They go by the appropriate name of near-Earth objects (NEOs).

Just a quick definition of the terms used in the following description: aphelion is the most distant part of an orbit from the Sun, perihelion is the closest, and 1 AU (Astronomical Unit) is the average distance of Earth from the Sun, roughly 149.6 million km (93 million mi.).

The NEOs are divided into four groups: Apollo, Amor, Aten and Atira. There are also dynamical classes of NEOs that can contain two or more of these. One, called Ajuna, consists of small bodies that are repeatedly trapped in a 1:1 resonance with Earth; only about

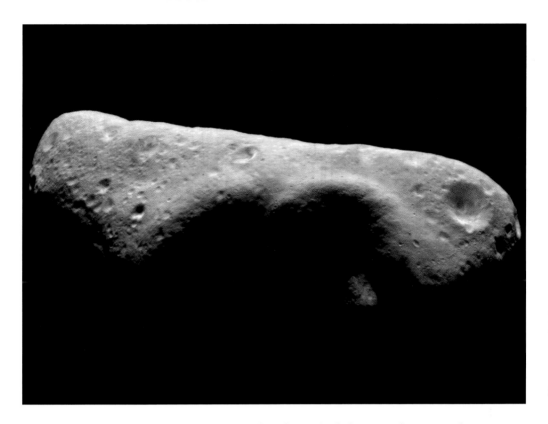

The near-Earth asteroid Eros as imaged by the NEAR spacecraft. This mosaic of four frames, photographed on 26 September 2000, was taken as the spacecraft looked down on the 'saddle' region, a broad, curved depression that stretches vertically across the image.

five are known. Another dynamical class, Earth-grazers, have a perihelion between Earth's perihelion and aphelion. Earth-grazers that do not get closer to the Sun than Earth are Amors, while those that do are Apollos. Yet another dynamical class that contains asteroids that can have perihelia very close to Earth's orbit are Alindas, named after its main asteroid 887 Alinda, discovered by Wolf in 1918. These asteroids, numbering 24 as of 2020, are held in a region of space defined by a 1:3 orbital resonance with Jupiter, which is also a 4:1 resonance with Earth. Thus every four years Earth has close encounters with some Alindas, including 4179 Toutatis, which is not only an Alinda but an Apollo asteroid. It came within 18 lunar distances of Earth (0.05 AU) in 2012; an even closer approach of 0.02 AU will happen in 2069. A Chinese spacecraft made a fly-by of Toutatis on 13 December 2012 at a distance of 3.2 km, but its main claim to fame is in our popular culture, as will be explained in Chapter Four.

Eros is classed as an Amor, the first one ever found. The class takes its name from 1221 Amor, discovered in 1932 and named after the Roman equivalent of the Greek god Eros. (Many asteroids are named for Greek deities, while others bear equivalent names from the Roman pantheon.) The regular change of brightness of an asteroid, caused by its rotation, results in a so-called lightcurve. 433 Eros has the distinction of being the first asteroid to have its lightcurve measured, in 1901.

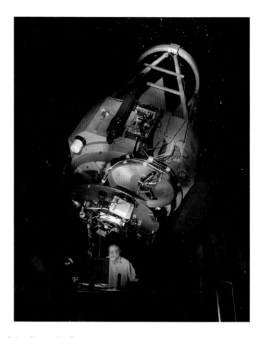

The 2.3-metre (90 in.) Bok telescope is operated by Steward Observatory on Kitt Peak in Arizona. It is named in honour of the astronomer Bart Bok (1906–1983).

Amors are defined by three criteria: their orbital periods are greater than a year, their orbits do not cross that of Earth (although some cross the orbit of Jupiter), and as part of the NEOs they must have a perihelion distance less than 1.3 AU. The population of Amors as of August 2020 is 8,551, but only 1,153 are numbered, which indicates that we have sufficient data to establish their orbits. In February 2020 the tiny Amor asteroid 2020 CD_3 became only the second-known asteroid to be in temporary orbit around Earth, the first being found in 2006. Such objects have been unofficially dubbed mini-moons.

Apollo asteroids number 12,860 as of August 2020, of which just over 1,400 have been assigned numerical designations. Their total population is estimated at 80 million objects the size of a house or larger. To be classed as an Apollo, an object must have a semi-major axis greater than Earth's, but a perihelion distance less than Earth's aphelion distance. The largest member of the group is 1866 Sisyphus at 8.5 km (5.3 mi.).

As of 2020 some 1,758 Aten asteroids are known, of which only a dozen are named. Their orbits are constrained by having a semi-major axis less than that of Earth, clearly demarcating them from Apollos. Further, their aphelion must be greater than 0.983 AU. Here is a description by Roy Tucker of his discovery of an Aten with the Steward Observatory's 90-inch Bok telescope on Kitt Peak in Arizona. Teamed up with a couple of professional astronomers, he was the first amateur to find an Aten.

During our one-week observing period in June 2004, we found a fast-moving object and reported positions to the Minor Planet Center. It was assigned preliminary designation 2004 MN$_4$. It wasn't until December of that year when it was recovered that it was identified as a possible Earth impactor in 2036. At one point that month, the probability was as high as 2.5 per cent [a 1-in-37 chance]. It was indeed a very exciting period. This Aten asteroid was later designated (99942) Apophis.

Artist's concept of how the break-up of asteroids around the Sun or other stars (exo-asteroids) generates debris and dust.

On 13 April 2029, Apophis will get very close to Earth: just 33,000 km (19,000 mi.) away, bright enough to be seen with the unaided eye. Owing to the publicity surrounding its close approaches to Earth, the 270-metre-long (885 ft) Apophis is now the poster child for risky asteroids, even though orbital calculations reveal it not to be an imminent threat to Earth.

Atira asteroids never have an aphelion greater than 0.983, meaning they never cross Earth's orbit, leading some to designate these as Inner Earth Objects. Only 23 Atiras are known; the class's largest member at roughly 4,800 m (3 mi.) is 163693 Atira, discovered in 2003. There is a short window each day in which one can find Atiras, only 20 to 30 minutes before sunrise or after sunset. A camera called the Zwicky Transient Facility (ZTF) at Palomar Observatory in California found two such objects in 2019 (each about 1,200 m (3,900 ft) across), just a year after it began operation. 'Both of the large Atira asteroids that were found by ZTF orbit well outside the plane of the solar system. This suggests that sometime in the past they were flung out of the plane of the solar system because they came too close to Venus or Mercury,' said Tom Prince of the California Institute of Technology. An especially exciting discovery was made on 4 January 2020, when ZTF found the first asteroid (2020 AV_2) with an orbit entirely inside that of Venus. Dubbed a Vatira as the first of a new sub-class of Atira, it has the shortest perihelion distance (0.7737 AU) of any object known. 'For NEOs we are finding objects that are small and moving quickly, between magnitude 24.5 and 28,' said Bryce Bolin of the ZTF. 'It fills a niche as we can see 10m-size objects. Only 1.5 per cent of our observing time is devoted to the twilight survey for Atiras.'

The idea of asteroids close to the Sun has a long history. An anonymous writer in 1860 stated, 'The calculations of Leverrier, and the discovery of Lescorbault, render it very probable that there is a zone of planets or asteroids near the sun.' A startling discovery about dust was announced in March 2019 that has a bearing on this idea. It has been known for many years that both Earth and Venus are escorted in their orbits by a band of cosmic matter, but data from NASA's STEREO satellite has revealed that Mercury too wades through dust. It was thought that solar radiation and magnetic forces would simply blow away any dust that close to the Sun, but researchers found a ring extending throughout Mercury's orbit nearly 15 million km (9.3 million mi.) wide. This dust originates from crumbling asteroids and comets, providing further clues about the formation of solar systems.[6] As for Venus, the dust in its orbit comes from a group of previously unseen asteroids co-orbiting with Venus. Evidence suggests that this population of crumbling asteroids has been there since the early formation of the solar system. It has been calculated that 8 per cent of the original population of asteroids in a 1:1 resonance with Venus still exist, but as of 2019 only five are known.[7] It is estimated that 3 per cent of these resonant asteroids can become Earth-crossers, thus posing a potential collision threat with our planet. The evolution of asteroids closer to the Sun than Earth is a rich area for exploration, including the hypothetical 'vulcanoids' that orbit closer to the Sun than Mercury.

Of the current population of NEOs – more than 23,302 as of August 2020 – we only have well-known sizes for 2,000 and rotational rates for a few hundred. 'The most accurate technique for NEO spectral type is near-infrared spectroscopy,' said Vishnu Reddy, 'but we are limited to magnitude 19.5 in the infrared.' Thus the composition of only the largest NEOs is known. It is not the asteroids we know that are a potential impact problem, but the ones we don't know. Leslie Mullen is host of the NASA podcast On a Mission. In the episode 'The Sky Is Falling' of 8 October 2019, she said, 'The small rocks from space do little to no damage. It's the big ones we worry about. And there are big asteroids out there, monsters lurking in the dark, large enough to take us out.' There is as yet no scientific consensus on the likelihood of a significant asteroid impact as it depends on these unknowns, including objects hidden from our view in meteor streams. But the threat is clear. 'The truth is', said Mullen, 'our planet is in the middle of a cosmic shooting gallery.'

Trojans

This type of asteroid, named after heroes of the Trojan War of around 1400 BCE, as chronicled by Homer in the *Iliad*, can theoretically be found in any planetary orbit. 2010 TK$_7$ has the distinction of being a lone wolf – it is the only known Earth Trojan, and is somewhere between 150 and 500 m (490 and 1,640 ft) in size.

No matter what planetary orbit they inhabit, Trojan asteroids occupy just two very special regions of space known as Lagrangian points, denoted as L4 and L5. Lagrangian points are named after the French mathematician Joseph-Louis Lagrange, who mathematically discovered them in 1772. Any object at these points, 60 degrees behind or ahead of the planet in its orbit, will maintain its position there (although it can oscillate many degrees from the exact point). While we know of only one Earth Trojan, it is estimated that Jupiter Trojans number in the millions. The first Trojan ever found was 588 Achilles in 1906, at the L4 point of the Jovian orbit. It is 130 km (80 mi.) across, but another Jupiter Trojan, 624 Hektor, holds pride of place as the largest, with an elongated body 403 km (250 mi.) long but only 201 km (125 mi.) around. It was discovered by August Kopff in 1907, but its size and shape were not known until the 1970s when the foundation of our knowledge of Hektor was laid by Dale Cruikshank and Bill Hartmann, using observations made with spectrometers built by Cruikshank at the University of Hawaii. Hartmann related to me an amusing anecdote from those days.

In the observing room at Mauna Kea, Hawaii, the telescope operator would be responsible for operating the telescope; Dale would be monitoring the spectrometer and data coming in moment by moment; so I used to start drafting the proposed

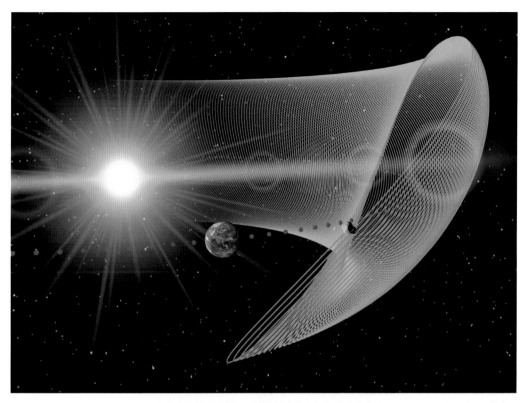

The orbit of 2010 TK$_7$, the first Earth Trojan asteroid, is denoted by green dots. Asteroid 2010 TK$_7$ has an extreme orbit that takes the asteroid far above and below the plane of Earth's orbit. In addition, the asteroid moves within the plane of Earth's orbit in what is called libration, circling horizontally around its stable point every 395 years.

William Hartmann (left) and Dale Cruikshank on an asteroid-observing run with the Infrared Telescope Facility on Mauna Kea, Hawaii, 5 April 1979.

multi-author paper – which seemed normal to me, since the first part of any paper was to explain the goals of the observing program and the background of the problem. But we (or everybody else who came into the observing room) used to joke, 'There's Hartmann over there, writing the paper before we have any results!' (Of course, in spite of cloudy nights, I always maintained the faith that Dale and our Hawaii colleagues would ultimately get something worth reporting.)

Telescopic observations in 2011 confirm that Hektor has a moon, the only known binary Trojan in the L4 point. We will encounter another binary Trojan, Patroclus (discovered in 1906), later in this book.

Trojans have been found in the orbits of five planets besides Earth. The dates associated with each are the first discoveries, followed by the number known as of 2019: Neptune (2010; 22), Uranus (2013; 2), Jupiter (1906; 7040), Mars (1990; 7); and Venus (2013; 1). Uranus may also have another connection with asteroids: its epsilon ring is composed of rocks, not dust-sized particles as other planetary rings are. These rocks may derive from shattered asteroids.

A key factor in Trojan asteroid research is their respective numbers at L4 (4601) and L5 (2439). This disparity has been recognized for decades, but only in 2019 was a theory, discussed later in this chapter, advanced to explain why there are so many more at L4, the trailing point in Jupiter's orbit.

Hildas

Occupying a zone just beyond the main belt, and extending to Jupiter's orbit, are the Hilda asteroids, named after its largest member, 153 Hilda. The 175-kilometre (109 mi.) Hilda was discovered by Palisa in 1875, the first of a group now numbering more than 4,000. The great majority of the population (except for the captured bodies such as comets) have origins tied to the formation of Jupiter. Asteroids given the taxonomic classes P and D (explained below) predominate among the Hildas; both have a low albedo of 0.05, meaning only 5 per cent of incident sunlight is reflected by these dark objects. P types are rare, numbering only 33; one of these is Hilda. Spectrally they look like M type (metallic) asteroids, but are differentiated by their albedo, which for M types is much higher (0.07 to 0.21). D types exhibit a very red spectrum beyond 0.7 microns and exist primarily among the Hilda, Centaur and Trojan asteroids; only 46 have been identified. Both P and D types have been linked to primitive meteorites known as carbonaceous chondrites, the most famous of which, the Allende meteorite, fell in Mexico in 1969.

Centaurs

The first Centaur, 2060 Chiron, was found in 1977 but it was not until 1992 that the second was discovered. Their population is estimated to exceed 44,000 with sizes larger than 1 km, but only about 450 have been observed. These are truly extraordinary objects as they represent a transition from Kuiper Belt objects to comets: they begin as inactive, solid, asteroid-like bodies sent in from the trans-Neptunian region. Chiron's secular brightness variation was first reported by Robert Marcialis and B. J. Buratti in mid-1987 and a search for pre-discovery images revealed another outburst in the 1970s; by February 1988 it had brightened by 75 per cent.[8] Subsequent study by Marcialis and Steve Larson in 1992 showed Chiron to be a comet with a full-fledged tail, so this object, about 181 km (112 mi.) in size as determined by a stellar occultation in 1993, is designated as both a minor planet and a comet (designated 95P/Chiron). While only two other Centaurs have been seen to exhibit comas of dust and gas, like a comet, our current understanding is that any Centaur that approaches the Sun

This artist's impression shows how Chariklo's rings might look from close to the surface.

A painting of 5145 Pholus and our distant Sun, by William Hartmann. This work dates from 1997, five years after its discovery by the Spacewatch telescope. Pholus is nearly as red as Mars, but probably coloured by organic molecules (non-biological), rather than Mars's rusted iron minerals.

by having its orbit perturbed will behave like a comet.

The extraordinary nature of Centaurs does not end there. The largest one known, 10199 Chariklo, was the first asteroid discovered to have two rings. It was found in 1997, but its rings were not known until a stellar occultation revealed their presence. At an estimated 232 km (144 mi.), the object is sufficently large for astronomers to consider it for membership in the list of dwarf planets (see below for a discussion of this category). The Centaur 5145 Pholus is one of the reddest known asteroids in the outer solar system.

The definition of a Centaur is not written in stone, but generally they have a perihelion beyond the orbit of Jupiter and a semi-major axis that is less than that of Neptune. Unlike main belt objects, Centaurs are in unstable orbits: very few of them remain in the realm of the giant planets for more than 1 per cent of the age of the solar system. Some Centaurs were identified in 2020 as interstellar objects.

Active Asteroids

Like Chiron, active asteroids straddle the line between comets and asteroids. If one regards Chiron primarily as a comet, the first active asteroid observed was 7968 Elst-Pizarro in 1996. Once it started to exhibit a tail, it was also listed as a comet with the designation 133P/Elst-Pizzaro, where the P stands for periodic. Even asteroids discovered long ago can join the active list after being studied with modern instruments. 596 Schelia was found in 1906, but it was not until 2010 that its coma was seen. Theorists are of the opinion that it was hit by a 35-metre-wide (115 ft) asteroid, triggering ejecta from the surface which were observed as a comet-like coma. Several other asteroidal objects seen by the Hubble Space Telescope also appear to be giving off gases like a comet. An active object that at first appeared to be a comet was identified in November 2019 as an asteroid (between 200 and 400 m in size) that was struck by a smaller asteroid in 2016. P/2016 G1, discovered by Richard Wainscoat and Robert Weryk, has since completely disintegrated.[9]

The most striking example of an asteroid with a comet-like tail is 6478 Gault. About 3.7 km (2.3 mi.) in size, it is a rare example of a self-destructing asteroid. Gault's period of rotation has been accelerating for perhaps 100 million years. The most likely cause is sunlight bouncing off the asteroid at different angles (released as infrared radiation), an effect that has been given the acronym

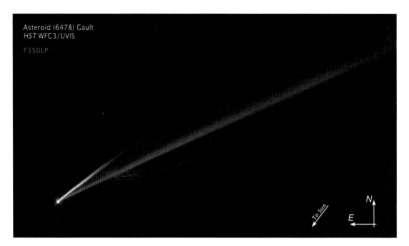

Asteroid (6478) Gault
HST WFC3/UVIS

F350LP

The self-destruction of asteroid Gault is shown in this image released on 28 March 2019. The longer tail stretches more than 800,000 km (500,000 mi.) and is roughly 4,800 km (3,000 mi.) wide. The shorter tail is about a quarter as long.

YORP, based on the surnames of four scientists whose work led to our understanding of it (Yarkovsky-O'Keefe-Radzievskii-Paddack). The effect is very slow: about 1 second every 10,000 years for Gault. Close study of asteroids subject to YORP can inform our geological knowledge: for example, landslides may occur due to faster rotation periods over long periods of time. In the case of Gault, the process may have reached a critical point, with the asteroid spinning every two hours. Substantial landslides causing dust ejections may be responsible for the twin tails. Sunlight is illuminating the material it is trailing, creating a stunningly beautiful two-tailed asteroid. The first tail was spotted on 5 January 2019 and the second in mid-January, but pre-discovery photographs show it had been active at least as far back as 2013. The asteroid may self-destruct if the process continues, but the length of its sustained activity suggests it may be a new type of object.

Trans-Neptunian Asteroids

The most distant asteroids in the solar system have a greater average distance from the Sun than Neptune (30 AU) and are thus known as trans-Neptunian objects (TNOs). Because most discoveries in this region are very recent, the terminology to denote them is a bit confused. This is exemplified by the object known as Pluto. It has variously been described as a planet, dwarf planet, asteroid and Kuiper Belt object (KBO). To codify its status as a dwarf planet, it was given an asteroid number by the Minor Planet Center in 2006: 134340.

Ever since the discovery of Pluto in 1930, there has been speculation about even more distant objects. The Kuiper Belt is named after the famed planetary scientist Gerard Kuiper, who speculated in a 1951 publication about the existence of solar system objects beyond Pluto that may have existed in the early solar system. Even though he did not believe such objects still existed, his name became attached to asteroids and comets now defined as having low-inclination orbits between 30 and 50 AU from the Sun. (Such objects thus form a subset of TNOs, a term that covers any object with an average distance greater than 30 AU.)

Kat Volk at the University of Arizona is at the forefront of research into TNOs. 'The current orbital structure of the trans-Neptunian region holds the key to understanding how the giant planets migrated

Comparison of the largest TNOs: Pluto, Eris, Haumea, Makemake, Gonggong, Quaoar, Sedna, 2002 MS$_4$, Orcus and Salacia. All except two of these TNOs (Sedna and 2002 MS$_4$) are known to have moon(s). The top four are IAU-accepted dwarf planets while the bottom six are candidates that are accepted as dwarf planets by several astronomers.

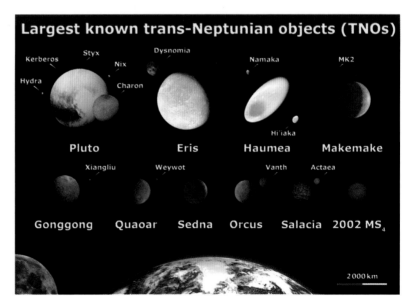

Largest known trans-Neptunian objects (TNOs)

Kerberos · Styx · Nix · Dysnomia · Namaka · MK2 · Hydra · Charon · Hiʻiaka

Pluto · Eris · Haumea · Makemake

Xiangliu · Weywot · Vanth · Actaea

Gonggong · Quaoar · Sedna · Orcus · Salacia · 2002 MS$_4$

2000 km

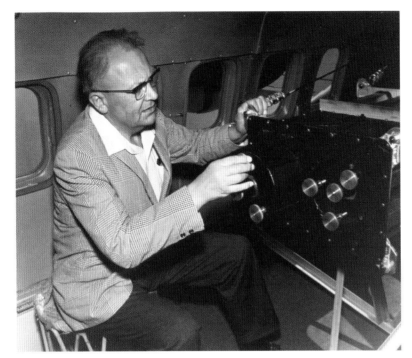

Planetary scientist Gerard Kuiper (1905–1973), after whom the Kuiper Belt is named, is shown here in an airborne observatory.

to their present-day orbits,' she states. Volk has identified upwards of thirty resonances, with an estimated population of tens of thousands of TNOs. One such resonance is at 9:1, where two objects have been identified, 130 AU more distant than Neptune. Her dynamical study indicates that these particular objects will stay there for a billion years. The physical study of TNOs is also yielding results. A study of them beyond 40 AU reveals they 'have distinct surface colours. We don't know what that means – they may have different surface ices,' Volk said at a University of Arizona seminar in October 2019.

As of 2020, more than 2,700 TNOs have been found, 2,000 of which have not yet received numbers, indicating that there are insufficient data to determine their orbits. KBOs are just one of the five sub-classes of TNOs. Another sub-class has been termed Sednoid, only three of which are confirmed as of 2020. Sednoids are defined as objects with perihelion greater than 50 AU and a semi-major axis greater than 150 AU. This sub-class takes its name from Sedna, an object 1,000 km (620 mi.) in size that was found in 2003; computer simulations suggest there may be forty Sednoids similar in size to Sedna that are yet to be seen. (Without getting bogged down in details, the other subcategories are resonant objects, scattered-disc objects (SDOS) and detached objects. The boundaries are fluid: Sedna is a sednoid, a scattered-disc and a detached object.)

If you think astronomers are getting into a muddle with all this terminology you may not be too far wrong. It is a manifestation in science of the 'enveloping category of categories . . . recklessly exceeding the requirement of the plot' described by Malcolm Bowie in his literary study *Proust Among the Stars* (1998). As we have just begun to read the book on the discovery of distant asteroids, I expect it will get properly sorted out as the plot becomes evident.

To describe objects that never come closer to the Sun than Neptune, even at its aphelion distance 30.33 AU, astronomers have created another category with the unwieldy title Extreme Trans-Neptunian Object, or ETNO for short. About twenty are known. This is where exciting discoveries are now being made. Juliette Baker, who has been at the forefront of studying these objects, related some of this to me in late 2018 when we met at the University of Texas in Austin. In 2014 an ETNO dubbed Deedee was found. It has a diameter of 630 km (390 mi.), making it a mid-sized dwarf planet. At its closest approach to the Sun, it is 37.97 AU away, but its average distance is 108 AU.

The most extreme of these extreme objects was found in 2015, and is designated BP_{519}. 'It is the first of a new class which will allow constraints on the process of giant planet formation,' explained Baker. 'It represents a new regime in the outer solar system.' With an estimated diameter as large as 700 km, it will likely also attain the status of a dwarf planet when we have more precise physical data.

While the inclination of the ETNOs is typically 30°, BP_{519} is inclined 54.1° and has a high eccentricity of 0.92, a highly elliptical orbit. The eccentricity of Neptune is just 0.009, a nearly circular orbit. Based on our current models of the solar system, 'it is difficult to explain its existence,' said Baker. One possible explanation is the presence of a ninth planet, a large object lurking on the edge of the solar system (maybe 700 to 1,000 times further from the Sun than Earth) that would produce objects like BP_{519} naturally. The possibility of a ninth planet has been on the mind of astronomers ever since the late nineteenth century. The discovery of Pluto in 1930 was hailed as that long-sought object, but its mass is too low to strongly affect any TNOs, including ETNOs. For many astronomers, including Mike Brown and Konstantin Batygin, the ninth planet is the Holy Grail of astronomy. It may be found using advanced telescopes of the twenty-first century, with a study of the orbits of very distant asteroids (including KBOs, TNOs and dwarf planets, all of which possess an asteroid number) that may be gravitationally altered by the ninth planet leading the way. That orbital clue may help astronomers search in particular areas of the sky. However, a study released in March 2019 by Antranik Sefilian and Jihad Touma posits not a ninth planet but the combined gravitational force of smaller TNOs combined with a huge disc of debris past Neptune's orbit.[10] This model accounts for observed eccentric orbits in some TNOs, which is being used by various astronomers as an indicator of a ninth planet. To accomplish the task, the mass of Kuiper Belt objects needs to be several Earth masses, not the 4 to 10 per cent of Earth's mass it currently appears to be. This suggests that there is a lot more material out there than has been detected, and it may be true that both a disc of material *and* the ninth planet are awaiting discovery.

One such object that appears to be gravitationally influenced by a distant planet or debris disc made a big splash in the news in late 2018. In October, just in time for Halloween, astronomers announced a study of the 'Goblin' planet, which had been discovered in October 2015. Proving that purple prose about asteroids was not confined to the nineteenth century, *Astronomy*

The largest Kuiper Belt object, Pluto, shown here in enhanced colour to highlight different surface compositions. Photos combined to create this image were taken by the New Horizons spacecraft in July 2015.

magazine began its story about the discovery by stating that the 'outer solar system is a dark, cold place'. Even though the name evokes Halloween, one can be assured that this object did not make its first appearance 'on a dark and stormy night', to quote the classic line by Edward Bulwer-Lytton. At perihelion it is 65 AU from the Sun, but aphelion takes it out to 2,300 AU. Only Sedna (at 76 AU) and another sednoid, 2012 VP113 (at 80.4 AU), have more distant perihelia than the Goblin, and only two others have a greater aphelion distance. It is likely some 300 km (180 mi.) in diameter, and is thought to be spherical, raising it to dwarf planet status.

As for objects whose orbits have been determined, the existence of super-distant asteroids has been confirmed: the aphelion of BP_{519} is a mere 862 AU compared to the current record-holder, 2017 MB_7, at an astounding 8,100 AU. This asteroid also has the highest eccentricity ever measured, 0.9989, taking it very close to a hyperbolic orbit. 2017 MB_7 also has the distinction of holding yet a third record, the longest orbital period of any asteroid, estimated at 260,000 years. When one considers Pluto has an aphelion of just 49 AU, an eccentricity of 0.2488 and an orbital period of 248 years, one can readily see we are now dealing with an entirely new and dynamic solar system populated by a host of distant asteroids, some of which are large enough to be deemed dwarf planets.

Of this host, only one has been visited by a spacecraft, 486958 Arrokoth. The spacecraft was New Horizons, which flew by Pluto in 2014. The relative importance of these two fly-bys was succinctly put in 2019 by Chris Lintott, professor of astrophysics at the University of Oxford and presenter of the TV programme *The Sky at Night*. 'Studying Pluto tells you about Pluto. Studying Arrokoth might yield a greater prize in unlocking the secrets of how planets, in general, grow.' Lintott evocatively describes Arrokoth as 'part of the rubble of a construction project which finished more than 4 billion years ago'.[11] New Horizons flew to within 3,540 km (2,200 mi.) of Arrokoth. The encounter took place at a distance of 6.6 billion km (4.1 billion mi.) from Earth on 1 January 2019.

Arrokoth is a contact binary: two formerly independent asteroids that became one when they came in contact with one another. Their combined dimensions measure 31 km (20 mi.) long and 19 km (12 mi.) wide at the broadest point. The larger of the two, dubbed Ultima, is three times the size of the smaller one,

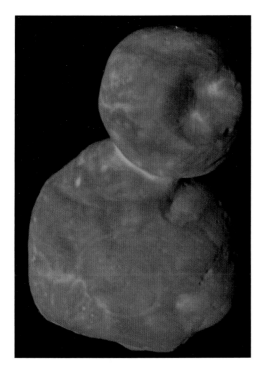

Arrokoth, the most distant object visited by a spacecraft as of 2021. The image combines enhanced colour data (close to what the human eye would see) with detailed high-resolution panchromatic pictures by the New Horizons spacecraft on 1 January 2019.

Thule. The first images received indicated both were spherical, but imagery obtained in February 2019 (which was taken 10 minutes after the closest approach) reveals something much more interesting. Ultima resembles a pancake, while Thule resembles a walnut. 'We've never seen something like this orbiting the Sun,' said Alan Stern, principal investigator on the New Horizons probe. Discovering that this asteroid is a contact binary excited astronomers around the world. Mark Buie, who discovered this asteroid in 2014, said in January 2019 when the first images were sent to us by the spacecraft, 'I looked at it and thought, "I see accretion happening." This is going to revolutionize our view of where we came from and how this whole process works.'

Scientific data revealed that the rotational axes of the lobes are nearly parallel, indicating that they formed out of one rotating cloud of icy particles in the protoplanetary disc at the beginning of the solar system. It is estimated that they came together at a speed of 2 m per second. Lines on the surface appear to delineate distinct clumps: these similar-sized objects may have merged together over a shorter time interval than the typical scenario of planetesimal growth, which is a long-term process. 'Just like Ultima Thule, they came together to form something wonderful,' said Kirby Runyon of the mission science team. Since the 1960s the prevailing theory has been that violent collisions formed the planets, but the study of Arrokoth has overturned that, confirming the gentle clumping theory proposed by Anders Johansen in 2005. In 2020 Alan Stern of the Arrokoth science team described this as a discovery of 'stupendous magnitude'.[12]

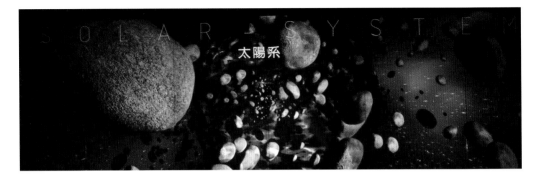

So what are planetesimals? The definition has changed over the years. In the 1980s astronomers regarded objects between 1 m and 100 m in size, which eventually (up to a million years) accreted to form larger objects, as planetesimals. Now the minimum size is set at 1 km, but the concept is the same: small objects in the solar nebula, during the formative years of the solar system, came together to form asteroids and planets. Hannes Alfven, who won the Nobel Prize in Physics in 1970, wrote to me in 1986 that his interest in asteroids went back to 1942, 'when I demonstrated that the asteroids could not be due to one or more "exploded planets" but represented a "planetesimal state" in a slow formation of planets'.

Arrokoth, which orbits the Sun with a period of 293 years, is the first pristine planetesimal to be visited by a spacecraft. Its two lobes are joined by a collar of material that is lighter than the two lobes, which reflect only 7 per cent of sunlight, indicating either a different chemical composition or smaller grains that are more reflective. Overall the asteroid is dark reddish, as would be expected given that sunlight has irradiated its surface for billions of years, producing a variety of organic compounds known as tholins.[13]

Depiction of planetesimals at the origin of the solar system, from a Taiwanese film about the formation of the universe in an exhibit entitled 'Star-studded Sky'. A 2020 study of type IIE iron meteorites has revealed some planetesimals had a liquid metal core that generated a magnetic field as strong as Earth's.

Dwarf Planets

One unscientific characteristic of dwarf planets is that they are all extraordinarily fascinating objects. The IAU adopted a definition of a dwarf planet in 2006 that relies on two criteria. First, it must be in direct orbit around the Sun and not be a moon orbiting another body; and second, it must be massive enough to attain a spherical shape. The criteria do not define how large an object must be to attain this designation. By 2020 the tally of dwarf planets officially recognized by the IAU (each of which retains a number identifying it as an asteroid) stands at five (Ceres, Pluto, Haumea, Makemake and Eris), but twenty or more (including the main belt spherical asteroid 10 Hygiea, 434 km (270 mi.) in diameter) may be added as the IAU considers the matter further; the table given below lists the eight largest, with Pluto as the largest TNO and KBO, and Ceres as the largest main belt asteroid; figures given for all objects except Pluto and Ceres are only estimates.

Like Pluto, Haumea resides beyond the orbit of Neptune so it is also called a plutoid. Haumea first surprised us when, in 2005, astronomers discovered that it has two moons. (Gonggong, the first major solar system body to have a Chinese name, also has a moon.) The second big surprise came in 2017: until its ring was found that year, no other TNO was known to have a ring. It appears that Haumea is elongated because of its rapid rotation of 3.9 hours, so its dimensions are estimated at 1626 × 1446 × 940 km.

Asteroid number/name	Diameter in km
134340 Pluto	2,372
136199 Eris	2,326
136108 Haumea	1,500
136472 Makemake	1,430
225088 Gonggong	1,230
50000 Quaoar	1,110
90377 Sedna	995
1 Ceres	939

Makemake, Haumea and Quaoar are sometimes closer to the Sun than Neptune, but Eris is always beyond Neptune. When Eris was discovered by a team led by Mike Brown in 2005, its size initially led NASA to refer to it as the tenth planet. Eris holds the dubious honour of being the largest solar system object not yet imaged by a spacecraft. It is only slightly smaller than Pluto, but is 27 per cent more massive. Like Pluto it has at least one satellite, dubbed Dysnomia. Eris is about three times the distance from the Sun compared to Pluto, with an aphelion of 97.46 AU. Stellar occultation measurements of Eris in 2010 confirmed Pluto as the pre-eminent member of the TNOs, but only by 46 km (29 mi.). Eris is classed as a scattered-disc object, indicating it was scattered from the Kuiper Belt; its unusual orbit can actually bring it closer to the Sun than Pluto, although this will not happen again for eight hundred years.

So how far out can one of these dwarf planets be seen? An object appropriately nicknamed FarFarOut answered that question as of

Eris and its moon Dysnomia, as envisioned in this artists' conception. The surface of Eris is likely covered in frost. As on Pluto, this is in the form of methane ice, but unlike Pluto, which is reddish, the surface of Eris is white.

21 February 2019 when its discovery was announced. Until then the most distant objects seen were Eris at 96 AU, and Farout at 120 AU, but FarFarOut is now 140 AU from the Sun. The orbits of the two most distant objects cannot be determined without further data, but FarFarOut is about 400 km (250 mi.) in diameter, and Farout appears to be more than 500 km (310 mi.), with the pinkish hue associated with ice-rich objects. One of the astronomers making these discoveries, Scott Sheppard, said, 'It's exciting to be looking at sky that no one has ever imaged as deeply as we are. To paraphrase Forrest Gump, each image we take is like a box of chocolates – you never know what you're going to find.'[14]

Interstellar

'Oumuamua is strange in every way, beginning with its Hawaiian name, which means 'a scout from the deep distant past'. Gregory Laughlin, a lead researcher of this object, says 'the name is extraordinarily evocative', as befits the first interstellar asteroid ever seen. 'Oumuamua, on a hyperbolic orbit from some unknown star

Artists' concept of the interstellar object 'Oumuamua, discovered in 2017.

system, was discovered on 19 October 2017 while it was exiting our solar system. 'There were a frenzied series of observations including a heroic observation of a spectrum on October 24, but it didn't see any features of any kind.' Before discovery it got quite close to the Sun, only 0.25 AU, receiving sixteen times more energy from the Sun than we get on Earth. Despite the fact that it was baked by this close approach, no comet-like coma has been seen coming from 'Oumuamua. 'The Spitzer space telescope looked at it on 17 November 2017,' said Laughlin at an astronomy conference in early 2019, 'but it did not see a coma. At most only a minute amount of micron-size dust is coming off it, too little to detect.' But the story is not so simple. 'It is not moving on a Keplerian orbit. It appears it accelerated as it left the solar system,' explained Laughlin, implying the object is being propelled by an invisible jet of gas. To add to the mystery, 'Oumuamua's brightness varies by a factor of 15, which implies the cigar-shaped object estimated to be 100 m long is tumbling chaotically. 'We know of no other asteroid that varies by a factor of 15. This starts to become a little weird,' Laughlin said. 'There are lots of mysteries but nothing completely crazy.'

The saga of 'Oumuamua merely begins there, however. Based on its unusual acceleration after passing by the Sun, Harvard University professor Avi Loeb wrote in *Astrophysical Journal Letters* in 2018 that the object may be of alien origin. 'Considering an artificial origin, one possibility is that 'Oumuamua is a lightsail, floating in interstellar space as a debris from advanced technological equipment . . . Alternatively, a more exotic scenario is that 'Oumuamua may be a fully operational probe sent intentionally to Earth vicinity by an alien civilization.' In 1960 Ronald Bracewell at Stanford University proposed launching probes to nearby stars to search for signs of intelligence, so Loeb is suggesting an alien version of a Bracewell probe may have been sent to us. It has even been suggested that the asteroid belt is a natural choice for any settlement by visiting extraterrestrials.[15]

Over the Hedge cartoon of 'Oumuamua, suggesting that it might be an extraterrestrial visitor.

In an early 2019 interview with CNN, Loeb said, 'It looks nothing like the asteroids or comets we have seen before in the solar system. Unfortunately this special guest we had for dinner already left our house and it's out there on the dark street so we can't really see it anymore, but we can look for other guests that came from that foreign country.' A theory released in March 2019 attempts to explain how the acceleration of the object may be due entirely to the effects of solar radiation. The theory posits that there was enough light pressure on the Sun-facing side to cause a jet of water vapour to develop. As 'Oumuamua tumbled, the sides facing the Sun would, in this scenario, have continued to eject water vapour whenever sunlight struck it, thus accelerating the body.[16]

One theory of its origin harkens back to the planetary disruption hypothesis that was so prevalent in the nineteenth century as an explanation of the asteroids in our solar system. It has already been observed that one-quarter of white dwarfs contain metallic debris from the break-up of exo-asteroids (that is, asteroids orbiting other stars). The violent formation of a white dwarf star may generate the radiance of 10,000 suns; that radiance hitting asteroids would increase the YORP effect so much that they would spin asteroids very rapidly and break up.[17] Some of these shards, like 'Oumuamua, could be flung across the galaxy. Olbers would likely be pleased by this concept. Another theory suggests that it is a chunk of molecular hydrogen ice. Whatever the truth may be, 'Oumuamua is certainly the most unusual object ever to be given the designation of asteroid: the Minor Planet Centre, which operates at the Smithsonian Astrophysical Observatory under the auspices of the IAU, gave it the designation 1I/2017 U1, where I signifies interstellar.

'Oumuamua has raised the biggest controversy since Ceres and Pallas were discovered more than two hundred years ago. In October 2019 it received a companion of sorts: the first interstellar comet. It is designated Comet 2I/Borisov (where the I again signifies interstellar) in honour of its discoverer, Gennady Borisov, a citizen astronomer from Crimea. It has also been suggested that another asteroid with a Hawaiian name, 514107 Ka'epaoka'awela, has interplanetary origins. This asteroid, found in 2014, is in a 1:−1 resonance with Jupiter; the minus sign indicates that it has a retrograde orbit, the first example of its kind.

Families

In addition to the resonances I mentioned earlier, there is another force at work that determines the location of asteroids. Families of asteroids can be formed by two processes. The first is collisional: when two asteroids collide, one may excavate smaller objects from the large asteroid that was hit, or both may shatter. Many of these fragments will continue to orbit the Sun in the same area of space. The opposite of collisions are fissions, whereby a precursor object ejects matter. This in turn may be caused by a collision that imparted a rapid rotation to the precursor. Fissions may also occur when the precursor expels a secondary body that breaks up. Whatever the cause, fully one-third of main belt asteroids are members of families. Their existence was first recognized by William Henry Stanley Monck, a professor at Trinity College Dublin, in 1888, but the name most associated with them is that of the Japanese astronomer Kiyotsugu Hirayama.[18] He identified five families in 1923; some one hundred are now recognized ranging in size from

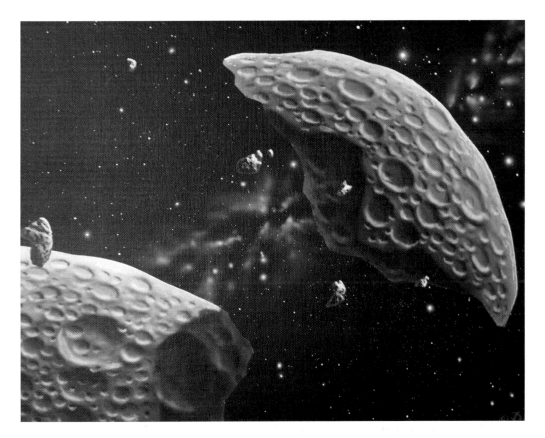

The collision of two asteroids as envisioned by space artist Kim Poor in 1984, in a painting entitled *Direct Hit*.

tens of members to thousands. They also exhibit a vast range of ages. The youngest are no more than a few million years old, while the oldest has been dated to 4 billion years.

Some families contain compositionally similar objects, while others have their membership assigned because of an orbital grouping. An example of the former is the Flora family, with more than 13,000 members, nearly all of which have reddish colours. This family, of which 8 Flora is the largest at 128 km (80 mi.), may have been created by the fragmentation of a binary or multiple object. Some 38 per cent of meteorites that land on Earth are a match for the type of objects found in this family, which is the likely source of those impacts on our planet.

65

With more than 19,000 members, the Nysa family is the largest. It is named after 44 Nysa, and like the Flora family is located in the inner region of the main asteroid belt. Unlike the Flora family it consists of two distinct mineralogical groupings. The asteroid Gault came from the Phocaea family, which consists of 2,000 objects in the inner belt region. In the outer asteroid belt we find the Koronis family of stony asteroids, with nearly 6,000 members, and the Hygiea family with nearly 7,000. Observations with the Very Large Telescope (VLT) in 2019 revealed a surprise: Hygiea (discovered in 1849) is spherical and lacks any large impact craters. The family was created 2 billion years ago when an asteroid between 75 and 150 km (47 and 93 mi.) in diameter hit a parent body that was completely shattered. Once most of the pieces reassembled, they formed a large spherical object with thousands of companion asteroids. Such a collision is unique in the last 3 to 4 billion years.[19]

Over the billions of years of solar system history, stray asteroids, known as interlopers, have also become mixed in with many family groupings. Thus making thorough studies of families is quite a complex task, but can reveal much about what happened in the asteroid belt in the distant past.

Like main belt objects, we now know that TNOs have families – that is, they exist in groups that share orbital characteristics. Haumea established its importance as the largest member of the first TNO family ever identified. Along with Haumea and its moons, the family consists of five other asteroids and was likely formed billions of years ago as the result of a collision.

Multiple Systems

Wherever asteroids are in the solar system, they have the potential to be attended by an orbiting moon. William Herschel was the first, in 1802, to suggest that asteroids might have smaller objects orbiting them, and he actively looked for moons orbiting Ceres and Pallas.

From then until 1993 no asteroids were known to be binary. As of 2020 the list of binaries has grown to an astonishing 390 asteroids with 370 known attendants.

How binary systems form is an active area of study.[20] One theory is that an asteroid, on a close approach to a gravitationally powerful planet such as Jupiter, can split in two. Patroclus is a good example of this type. Others, such as Arrokoth, are contact binaries, formed when two asteroids collide slowly and merge into one. Still others may be due to capture, in which a relatively large asteroid brings a smaller object into its gravitational embrace. Or a so-called rubble pile asteroid, which consists of a fairly loose agglomeration of rocks both large and small, may be hit. When this happens some of its structure may be loosened enough to dislodge a substantial chunk that does not have enough energy to escape. It stays close to its fellow members of the rubble pile, thus forming a binary system.

Artists' concept of the Trojan asteroid Patroclus and its moon Menoetius.

The first detection of a binary came not from a diligent Earth-bound observer, but from the Galileo spacecraft, which passed by 243 Ida on 28 August 1993 on its way to Jupiter. During this fly-by it photographed Ida and its moon, later dubbed Dactyl, which was revealed in images studied in February 1994. Ida (discovered in 1884 by Palisa) is a potato-shaped asteroid measuring 60 km (37 mi.) along its main axis with an average diameter of 31 km (19 mi.). Dactyl is puny by comparison, one-twentieth as big. Dactyl's origin is uncertain, but it may have been a fragment blasted off the surface of Ida in a collision within the last hundred million years.

Franck Marchis of the SETI Institute led a team that discovered the first triple asteroid system. The asteroid 87 Sylvia, discovered by Norman Pogson in 1866, was found to be a binary asteroid by Michael E. Brown and Jean-Luc Margot in 2001. In 2004 Marchis found a second moon orbiting Sylvia, which was named in honour of the legendary Rhea Silvia, mother of the founders of Rome, Romulus and Remus. These names were given to the two tiny moons of Sylvia: Remus is about 7 km (4 mi.) in size, while his big brother Romulus is 11 km (7 mi.).

Artist's rendering of the first known asteroid with two satellites: Sylvia and its moons Romulus and Remus. Fifteen more ternary systems have since been discovered.

All these are main belt objects, but TNOs also exhibit multiple systems. Pluto has five satellites, and its fellow dwarf planet Haumea has two. TNO satellite systems likely formed in the first 700 million years of the solar system from impacts between molten progenitors, according to a study released in June 2019 by scientists at the Tokyo Institute of Technology. We also know of Mars-crossing objects, Jupiter Trojans, Centaurs and near-Earth objects that are binary. (One such NEO passed by Earth at a distance of 5 million km (3 million mi.) on 25 May 2019.) As of 2020 there are 390 asteroids with known or suspected moons, including sixteen ternary systems, and this list is growing at the rate of one per year.

Migration of Planets and Asteroids

The biggest discovery of 2018 was announced in May, when a team led by Tom Seccull identified a Kuiper Belt object as an interloper asteroid. 'This is the first time we've actually seen an object that looks like it formed in the inner solar system. That's why it is such an important discovery,' Seccull explained. The object in

Artist's impression of the Kuiper Belt asteroid EW$_{95}$, thought to be an interloper from the inner solar system.

question was discovered in 2004, and bears the number 120216 and the provisional designation 2004 EW_{95}. It became apparent from reflectance spectra that EW_{95} was unusual, which prompted a detailed look with Europe's Very Large Telescope (VLT) in Chile.[21]

Just as a DNA test can reveal what grouping of humanity you have kinship with, a close look at the light reflected by these dark and distant KBOs can reveal their ancestry. And just as human blood groups can be distinguished by letters such as type A or type B, the asteroids are also grouped into letter classifications. In technical terms, the VLT revealed the presence of both ferric oxides and phyllosilicates (such as mica, widely used in manufacturing) in EW_{95}, two materials never before seen in a KBO (including Pluto), but common in carbonaceous asteroids in the main asteroid belt. Ferric (iron) oxides are the same sort of rust that gives Mars its reddish colour. This is formed by low-temperature (colder than 320K) chemical alteration of materials by liquid water. The water acts as a solvent to produce hydrated minerals such as ferric oxides and phyllosilicates, the other 'DNA strand' discovered on the 290-km EW_{95}. Phyllosilicates are minerals that consist of silicon and oxygen atoms arranged in sheets, and they all are attached to either water or some other form of oxygen bonded to hydrogen. The important point here is that liquid water was present on EW_{95}, probably melted by radioactive decay. This asteroid is classified as a plutino, a group of trans-Neptunian objects in resonance with Neptune: for every two orbits made by a plutino, Neptune makes three.

We know from a study of Ceres that it also possesses liquid water. Ceres is the largest of a group of 'primitive' asteroids denoted by the letter C, the so-called carbonaceous asteroids. Thus by discovering the two strands of 'mineral DNA', we realize at once that EW_{95} is not just an object in the Kuiper Belt, but an asteroid, one that is a member of the C-types. It must have formed in the outer asteroid belt (closer to Jupiter than Mars), as Ceres did, and it

The Very Large Telescope (VLT) on a mountain at a height of 2,635 m (8,645 ft) in Chile. It consists of four individual telescopes, each 8.2 m (27 ft) in diameter, that became operational between 1998 and 2000.

retains its so-called primitive nature because unlike asteroids that formed in the inner asteroid belt, EW_{95} and other C-types retained the fragile legacy of water bequeathed to them in the formation of the solar system. An investigation of two more such objects in similar orbits is now underway. 'We can use the orbits and characteristics of these objects to find out more about where they formed and how they moved as a result of the migrations of the giant planets,' said Seccull.

71

In astronomy it is possible to put forward a theory which will be generally acceptable, hold the field for a long period, and then be shattered by a new event, newly observed fact or newly developed theory. Such is the case with the positions of the planets and asteroids in relation to the Sun. The nebular hypothesis was generally accepted as explaining the presence and behaviour of asteroids. The idea was that the matter of which the planets are composed once floated in rings of gas and dust in the form of nebulae around the Sun. The material composing them was gradually drawn towards the densest part of each nebula, in time forming a single body. For the asteroids, it was thought that the material failed to do this, or perhaps there were too many more or less equally dense areas, causing the nebula to form into countless fragments. Until 1980 it was thought that both the planets and asteroids formed from the nebulae where they are currently located, but more recent dynamical modelling strongly suggests (but does not prove) that the giant planets migrated to their current locations. The most promising hypothesis that deals with this issue is dubbed the Grand Tack, but keep in mind that it was originally formulated to explain (among other things) the Late Heavy Bombardment (see below), evidence for which has been conclusively debunked.

The Grand Tack theory envisions Jupiter and Saturn sailing the dark seas of the heavens, rather than being fixed in their orbits. Because of this, the giant planets scatter about 15 per cent of the inner solar system planetesimals (assumed to be S-type objects, which are primarily stony and nickel-iron types) away from the Sun as they migrate into the inner solar system. (Jupiter 'tacks', in sailor's jargon, to just twice Earth's distance from the Sun.) As Jupiter and Saturn then migrate further away from the Sun, some 1 per cent of the S-type objects that were sent outwards are scattered back into the inner asteroid belt while some C-type objects, previously undisturbed, get scattered into the outer asteroid belt. As explained by Thomas Burbine of Mount Holyoke College, Grand

Tack 'seems to pull out the taxonomic distribution of asteroids in the asteroid belt where S-types are in the interior of the belt and C-types in the exterior (a gross simplification)'. While the observed ordering of S nearer the Sun and C in the outer main belt holds true for larger asteroids, it does not apply to small fry.

So how well does this hypothesis explain where the asteroids are? Research by Rogerio Deienno and colleagues shows that the Grand Tack model is compatible with the orbital distribution of asteroids we observe now. Their research paper begins with a supreme understatement: 'The Asteroid Belt is challenging to understand but is critical for studies of the formation and early evolution of the Solar System.' They find that the primordial eccentricities of Jupiter and Saturn have a major influence on the dynamical evolution of the asteroid belt. After the first 400 million years, the model indicates that three-quarters of the original mass of planetesimals remained in the main belt, some six times the mass we currently observe.[22]

Continuing the simulation for another 4.5 billion years, their study finds that about half that mass is lost. Even though the final result gives a mass of asteroids that is twice or even triple that which is currently observed, it is deemed to be a 'satisfactory match of the current Asteroid Belt mass'. While I do not regard such a disparity as satisfactory, it highlights the difficulty of arriving at a dynamical model that reflects reality as we see it.

In March 2019 a major challenge was mounted to a theory that I have already indicated is on shaky ground. Instead of having Jupiter form at 3.5 AU, then migrate to 1.5 AU and drift to its current position of 5.2 AU from the Sun, this proposal sees Jupiter forming four times further away from the Sun, starting out as an icy asteroid. The study, led by Simona Pirani at Lund University in Sweden, posits that after 2 to 3 million years gravitational forces from surrounding gas in the solar nebula pushed Jupiter inwards to its current position in a brief span of only 700,000 years. Key to the study was an attempt to explain the unusual distribution of Jupiter's Trojan asteroids, as detailed earlier in this chapter. That distribution is a consequence of having Jupiter migrate inwards: asteroids from 'the massive and orbitally eccentric Hilda group is captured during the migration' in a region between 5 and 8 AU, leading to the creation of the eccentricity and asymmetry we observe now. The model also predicts a total mass for the Trojans that is three to four times what is observed, so further understanding of the dynamics of the Trojans is required. The study suggests that Trojan asteroids have a composition similar to that of Jupiter's core.[23] Pirani further explains:

> Nowadays, we have many hints from hydrodynamical simulations and from some peculiar architectures in exoplanetary systems that planets migrate during the gas phase of the protoplanetary disc. When it comes to our Solar System it is hard to disentangle the events in the formative period. Luckily, Jupiter Trojans preserved this peculiar asymmetry that we have shown is a natural consequence of Jupiter's inward migration during its growth. The large scale inward migration of Jupiter also has effects on the asteroid belt. We have shown that a migrating Jupiter is able to implant outer solar system asteroids into the asteroid belt. This could be an explanation for the presence of P-type, D-type asteroids in the asteroid belt and also Ceres. In fact, due to its composition, Ceres is thought to have been formed in the outer solar system.

We cannot accurately plot positions of the planets for more than about 60 million years into the past or future owing to the fact that the solar system is chaotic. However, precise values for the perihelion of the planets have been found in the geological record. As the authors of a 2019 study show, extending our data back 223 million years has now given us the ability to constrain models of solar-system evolution. The work of mapping the chaotic evolution of the solar system even further back in time will be a major project for the next few decades.[24]

The most important aspect of the pinball solar system, as the ricocheting of the gas giants through the early solar system is popularly termed, is that icy objects from the outer belt may have been forced into the inner solar system, where they collided with Earth. Such objects could be the source of much of our water, which was essential to the development of life. It was determined in 2015 that water on our Moon was the result of asteroid (not comet) impacts. Obviously there is no liquid water on the surface, but chemically bound water on some asteroids would have survived a lunar impact. Such hydrated minerals are thought to populate as much as 4.5 per cent of lunar craters, and since such objects hit the Moon they certainly hit Earth as well, where conditions allowed that trapped water to be released. The flip side of asteroid collisions with Earth is their destructive effect, a topic examined in the next chapter.

THREE

EXPLOSIONS AND IMPACTS

Imagine our view of the cosmos in 1800 as a grand tapestry or woven image. The discovery of the first four asteroids unravelled it. The great textile artist Anni Albers wrote in 1938 that 'Anyone seeking to find a point of certainty amid the confusion of upset beliefs, and hoping to lay a foundation for a work which was oriented towards the future, had to start at the very beginning.' This was the meaning of her famous phrase 'the event of a thread': it represented an origin point. The discovery of the asteroids was the event of a thread, and it compelled astronomers to weave a new tapestry of the cosmos that depicts more about the origin of the solar system than any eighteenth-century researcher ever imagined. An actual complex textile web, as a collaboration of art and science to explore the cosmos, is currently being developed at the Art Institute of Chicago on basic principles established by Albers.[1]

Wilhelm Olbers, in a letter to Carl Gauss of 15 May 1802, sowed the seeds of an idea that would become one of the greatest controversies of the nineteenth century. 'I still can't wholly abandon the idea that Ceres and Pallas are maybe just fragments of a former planet.' Gauss realized at once the astonishing implications: nothing less than the ruin of the 'stability of the planetary system'. He could hardly gauge the shock this would give to the 'framework of knowledge' that underpinned that apparent but nonexistent stability. Instead of being built on bedrock, it

Left: Wilhelm Olbers, an astronomer in Bremen, suggested that asteroids were the remnants of an exploded planet. He discovered 2 Pallas (in 1802) and 4 Vesta (in 1807).

Right: Carl Gauss, whose mathematical genius was applied to the orbits of asteroids in the early 19th century.

would reveal instead that our understanding was 'built on sand, and that everything is entrusted to the *blind* and *random* play of the forces of nature!' This, in fact, is the very understanding of nature that modern research has revealed. Pallas proved to be the spark that lit the fuse, but for the wrong reason: Ceres and Pallas (along with the other asteroids) are *not* the result of the explosion of a primordial planet. Nonetheless nearly every printed description of the asteroids throughout the nineteenth century mentioned the possibility (often couched as a near certainty) that such an explosion happened. Olbers's hypothesis was a catastrophic success.[2]

This delusion, from the very beginning, became couched in religious terms. In 1802, Johann Bode asked, 'and why this considerable gap between Mars and Jupiter, where heretofore one finds no planet? Is it not likely that in this space wanders one of the celestial bodies to which the finger of God gave motion?' This phrase was used again in a text from 1850 to explain that the origin of asteroids was a planetary wreck.

> As we recognize the finger of God in one case, we must do so in the other. Though the normal state be stable, yet we find that through all God's works there are occasional disturbances of equilibrium. Such disturbances, by a natural extension of the word, may be designated *storms*. The solar system may be subject to its storms. We can conceive a cometary collision producing a cosmical storm, and astronomers imagine that they find the wreck of such a storm in the asteroids between Mars and Jupiter.[3]

In 1807 Johann Schroeter made observations that bolstered the idea that asteroids were fragments of an exploded planet. This is part of his report.

Johann Schroeter, who observed the asteroids with the largest telescopes in Continental Europe, mistakenly thought Ceres and Pallas were surrounded by huge atmospheres.

> I have observed Pallas this year whenever permitted by the weather, whereby I attended particularly to the variation of its light which cannot be missed in spite of the small size. The changes really seem to be periodical, and I therefore believe that they are not atmospheric but are rather caused by the rotation of the planet. It is strange, though, that the changes in light occurred so rapidly that they were remarkable already after 40 minutes. This is not unfavourable for Dr Olbers' hypothesis on their origin.[4]

Even though Pallas does not visibly change its light on time scales as short as 40 minutes, this report made it clear that asteroids are rotating objects, which is correct. In 1812 Lagrange was the first to apply mathematics to the hypothesis of Olbers, which, he wrote, 'extraordinary as it may appear, is not however improbable'. He calculated that if a planet had been torn asunder, its parts might form small planets orbiting the Sun. His work is actually more interesting for why it happened than for what he concluded. During the French Revolution, intellectuals were in danger. He was threatened with arrest and only escaped execution when a highly placed friend compelled him to make calculations on a subject of great importance to the regicides who had taken over the French government. The fate of the Revolution hinged on military power, so the theory of projectiles was paramount. How to aim a cannon to hit an enemy target depended on physics paired with mathematics. Years later, when Lagrange made his calculation about the asteroids, he explained that asteroids would go into solar orbit if they escaped the protoplanet with a velocity twenty times that of a cannonball.

Two writers in England, separated by 45 years, caught the true essence of the origin of the asteroids. In 1831 the scientific amateur William Coldwell wrote, 'Launched into [the] ether, these minute orbs have survived the rush of the ages equally with the larger spheres; yet do they seem to us sprung up yesterday, so completely have they for ages been hidden from us.' In 1876 the Cornish writer Nicholas Michell expressed it best poetically when he also rejected a planetary explosion for the origin of the asteroids. Like many of the Victorian era, he attributed to a deity what we now describe in terms of physics and chemistry, but nonetheless his lines are very evocative. Of the asteroids, he writes,

> . . . some have deemed
> The shining specks the ruins of a world
> Once huge and glorious, long, long ages back,

Shivered to atoms by convulsive throes,
While the torn fragments 'round the firmament
Must roam forlorn for ever. – Blind belief!
As if Omnipotence, in moulding worlds,
Could wander into error. Spirit, no!
With high design He cast these drops of gold,
These fairy orbs, along the aërial road.[5]

Asteroids have long been associated with gold and other precious metals. An 1896 science fiction novel by the Russian rocketry pioneer Konstantin Tsiolkovsky speculatively imagined the incredible mineral wealth of the asteroids. Even the small one his space travellers first land on has enough gold and platinum 'to pave streets'.

Formation of the Asteroids

The formation of the asteroids is intimately bound up with the formation of the solar system itself. How planets formed around the Sun has been the subject of serious scientific inquiry since the late 1700s when Laplace developed the nebular hypothesis, but even basic issues of planetary formation remain uncertain. When Thomas Chamberlin was writing in 1928, he quaintly attributed the formation of the 'planetoids' to eruptions from the Sun caused by a passing star.

A fundamental issue that needs addressing is how Earth itself was formed. In 1810 James Dean, professor of mathematics and natural philosophy at the University of Vermont, colourfully described the situation in classical terms. 'And, when with daring retrospection, we boldly attempt to describe the agitations of chaos during the organization of our planet, we draw a leaden blade against the shield of Ajax, and the distortion of our weapon must betray its insufficiency.' David Bennett, one of the panellists at a discussion about exoplanets during a conference on telescopes (held in Austin in June 2018), made a statement that many would find astonishing. 'The current state of affairs is that planetary formation theories have yet to make a prediction that has been confirmed.' While we certainly have achieved a greater understanding in the two hundred years since the asteroids were discovered, the comments of Dr Bennett are humbling, and must be borne in mind in any discussion of planetary or asteroid formation. Our weapons of study are still insufficient.

The panellist Nick Siegler explained that 'the most common type of planets [found around other stars] are between half Earth's mass, and twice its mass. This is extraordinary, it was not even predicted.' In our solar system, Earth is dwarfed by the planet Jupiter. 'Do stars with Earth-like planets typically come with Jupiter-size planets?' Scott Gaudi asked rhetorically at the panel. 'This is a crucial issue because delivery of water to the Earth in its early history was likely influenced by Jupiter.' This is why more information is needed about the stars. He stated a maxim that will govern this field of study for many years: 'Know thy star, know thy planet.'

Bennett said, 'if we really understood how planets form we would be able to put the information we get in context.' The search for that answer got a major boost just a month later in July 2018 when researchers at the Max Planck Institute for Astronomy in Germany announced that they had imaged, for the first time, a large planet in the process of formation. There is an interesting historical link involved.

The discovery was made with the VLT in Chile, which consists of four separate 8.2-metre (27 ft) telescopes whose light gathering can be combined electronically. Each of these large mirrors is made of Zerodur – the material of choice for the Extremely Large Telescope (ELT), too, which will begin observations around 2025 from its site in Chile using a 39-metre (128 ft) main mirror.

Both the VLT and the ELT will be crucial to improving our understanding of the formation of objects ranging from large

planets to asteroids around other stars, advancing the work begun by the Atacama Large Millimeter Array (ALMA), which began observing in 2011. Our first images of the surface of 2 Pallas came from the VLT. Released in February 2020, they reveal an extraordinary sight. 'Pallas' orbit implies very high-velocity impacts,' said the study's lead author Michaël Marsset.

A three-dimensional model of the European Southern Observatory's ELT in its enclosure. When completed in 2025, it will be the world's largest optical telescope, with 798 hexagonal segments forming a mirror 39 m (128 ft) in diameter.

> We can now say that Pallas is the most cratered object that we know of in the asteroid belt. It's like discovering a new world. Pallas experiences two to three times more collisions than Ceres or Vesta, and its tilted orbit is a straightforward explanation for the very weird surface that we don't see on either of the other two asteroids.[6]

Of the 4,000-plus exoplanets currently known, only the planet orbiting the T Tauri-type star PDS 70, 370 light years away, has

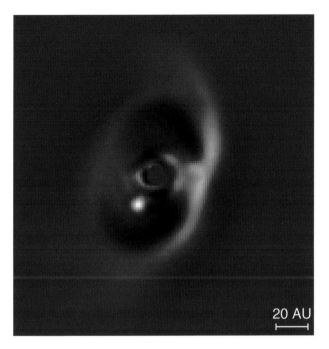

20 AU

The white dot is a planet orbiting the star PDS 70 with a 120-year orbital period. The dust and gas surrounding the star may eventually form an asteroid belt. This spectacular image taken in 2018 by ESO's Very Large Telescope is the first clear image of a planet caught in the act of formation.

been caught in the process of formation, but this 2018 discovery will certainly be followed by many more as larger telescopes and space-based observatories let us penetrate these systems with more detail. Are asteroid belts common around other stars, and if so, what type of stars? By the 2030s we should have some answers to these questions, although plotting the orbits of individual asteroids will be a project for the latter half of the twenty-first century. Exocomets have already been discovered (in 2018 and 2019) in observations by two space telescopes, and a planetesimal was spotted in 2019 orbiting a white dwarf star 410 light years away, so individual asteroids will be next.

Another major element in the formation of the asteroid puzzle is how the Kuiper Belt formed. Crucial to solving this puzzle is the size distribution of objects both now and in the distant past. In 2019 astronomers released a study of the cratering record on both Pluto and its satellite Charon. While the presence of craters larger than 10 km was expected, below that limit craters were much more scarce than predicted. This indicates fewer than expected KBOs smaller than 1 km in diameter, suggesting that most KBOs which are now extant (which are small) are primordial. The analysis also implies that whatever process led to the creation of KBOs tended to move masses into the creation of larger objects, whose former existence is evident from the cratering record.[7]

Late Heavy Bombardment

For the past 45 years an idea has held sway that lunar meteorite impact melts confirmed an era of intense asteroid/meteorite impacts on the Moon around 4 billion years ago (4.0 Gyr), with little to no impacts in the 400 million years before that. Despite a series of papers over many years by Bill Hartmann that questioned this belief, it became the canonical model, widely referred to as the LHB or Late Heavy Bombardment (late in the sense that it was more recent than the first 400 million years of the Moon's existence). By 2019 that model had collapsed under the weight of evidence to the contrary.[8]

Hartmann maintained that the newly accepted understanding by the astronomical community

> casts a whole new light on lunar samples, because the incredibly intense asteroid impact fluxes before 4.0 Gyr ago would have pulverized tens of kilometres of the crustal layers of the moon (which is my explanation for why we don't see the impact melts from that time), which would be created in the upper layers – but we can still see old 4.4 Gyr-old crustal rocks that are excavated by big recent craters like Copernicus, since they can penetrate through the megaregolith, at least in some regions. The asteroidal (and cometary) impact history was vastly more important to lunar history and our samples than realized before the Apollo and Luna landings, and even by many rock analysts after the Apollo and Luna missions! [The Luna missions were unmanned Russian probes that landed on the Moon.]

There is, however, some evidence that the impact rate on the Moon and Earth increased by a factor of 2.6 (compared to the rate going back to 1.0 Gyr) about 290 million years ago. Additional

84

evidence disproving the LHB comes from Mars: research released in June 2019 shows a massive asteroid impact older than 4.48 Gyr created Mars's hemispheric dichotomy, with no later cataclysmic bombardments.[9]

Extinction of the Dinosaurs

Dinosaurs and asteroids have become synonymous in the public imagination. When my first book about asteroids was published in 1988, the entire front cover (by artist Richard Bernard) was dominated by a wonderful image of dinosaurs, two of which were looking up at the skies to witness the asteroid that was about to wipe out 75 per cent of all species on Earth, including all the dinosaurs except birds; small mammals survived, which evolved over the next 65 million years to produce primates, including humans. Meanwhile debris from the impact travelled through the solar system, landing on Mars, Saturn's moon Titan, and Jupiter's moons Europa and Callisto, in the process perhaps seeding those objects with living microbes from Earth. Life may indeed exist on those moons, and it may have come from Earth.

The asteroid-extinction hypothesis was promulgated by Luis Alvarez, winner of two Nobel Prizes, who worked with me in writing the dinosaur chapter in my 1988 book. His hypothesis has received repeated confirmation over the ensuing decades, especially when the impact site itself (known as Chicxulub) was found in the Yucatán peninsula of Mexico. That was in 1991; in 2010 a consortium of 41 scientists gave it their final stamp of approval in a definitive edition of the journal *Science*, declaring that an asteroid impact was the cause of the extinction. While a few recalcitrant members of the scientific community still dissented, the matter was finally laid to rest in 2019 with the publication of a study by Robert DePalma and other scientists including Walter Alvarez, the son of Luis.[10] In the so-called Hell Creek geological formation in parts of the Dakotas,

Montana and Wyoming, he found fossilized remains from the very day of the asteroid impact. The most important paleontological discovery of the twenty-first century includes fallout in the form of microtektites that still contain glass from the day of impact. The presence of dinosaur bones within the incident site proves that the dinosaurs had not been decimated by other disasters en masse prior to the impact event, the first time dinosaur remains have been found in sediments laid down within a few thousand years prior to impact. This proves they did not all go extinct pre-impact, as some scientists have been contending, citing volcanic eruptions, climate change and other factors that were not caused by an asteroid.[11]

'Perhaps the most widely discussed aspect of asteroid/comet science in recent decades', writes Don Davis,

The asteroid impact generated a tsunami-like wave in an inland sea that killed and buried fish, mammals, insects and dinosaurs. Artist's conception courtesy of Robert DePalma.

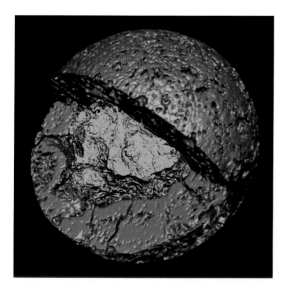

Micro-CT image showing cutaway of clay-altered ejecta spherule with internal core of unaltered impact glass that was formed when the dinosaur-killing asteroid hit Earth. This is from the Tanis site in North Dakota.

is whether or not one will again fall out of the sky and trigger an encore of the Chicxulub extinction – perhaps including humanity. The fact that asteroids and comets have hit the Earth, along with other solar system bodies, has long been known. But these were considered to be long-ago, rare events – not something that we have to worry about today. However, increased discoveries of bodies whose orbits cross that of Earth caused scientists to address the impact threat seriously. NASA is currently exploring what we can do if an asteroid is discovered that is on a path to impact the Earth.

Threat to Human Life

In April 2019 NASA administrator Jim Bridenstine said, 'Dinosaurs did not have a space programme. But we do, and we need to use it.' NASA classifies any NEO larger than 140 m (460 ft) with a minimum approach distance less than 7.4 million km (4.6 million mi.) as 'potentially hazardous'. But even smaller objects are a great concern. On 18 December 2018 Earth received a wake-up call that very dangerous objects can collide without warning. Exploding with ten times the energy released by the Hiroshima atomic bomb, a fireball several metres in size ended its life 25.6 km (16 mi.) above the Bering Sea. Something this big is expected about three times every century. An even larger object exploded over the Russian town of Chelyabinsk in 2013, causing a lot of property damage but no casualties. Having two such incidents in only six years suggests the expected frequency of such objects is higher than is

87

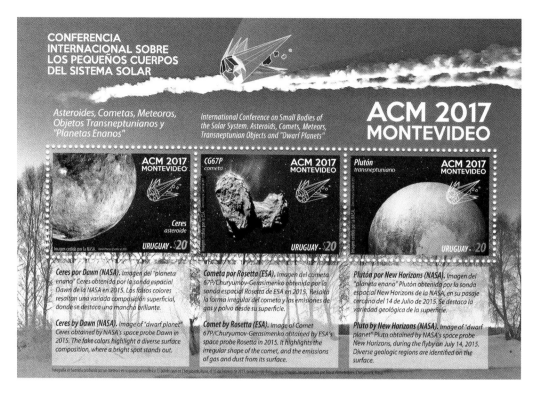

CONFERENCIA
INTERNACIONAL SOBRE
LOS PEQUEÑOS CUERPOS
DEL SISTEMA SOLAR

ACM 2017
MONTEVIDEO

Asteroides, Cometas, Meteoros,
Objetos Transneptunianos y
"Planetas Enanos"

International Conference on Small Bodies of
the Solar System. Asteroids, Comets, Meteors,
Transneptunian Objects and "Dwarf Planets"

ACM 2017
MONTEVIDEO
Ceres
asteroide
Imágenes cedida por la EUA
URUGUAY · $20

CG67P
cometa
ACM 2017
MONTEVIDEO
URUGUAY · $20

Plutón
transneptuniano
ACM 2017
MONTEVIDEO
URUGUAY · $20

Ceres por Dawn (NASA). Imagen del "planeta
enano" Ceres obtenida por la sonda espacial
Dawn de la NASA en 2015. Los falsos colores
resaltan una variada composición superficial,
donde se destaca una mancha brillante.

Ceres by Dawn (NASA). Image of "dwarf planet"
Ceres obtained by NASA's space probe Dawn in
2015. The fake colors highlight a diverse surface
composition, where a bright spot stands out.

Cometa por Rosetta (ESA). Imagen del cometa
67P/Churyumov-Gerasimenko obtenida por la
sonda espacial Rosetta de ESA en 2015. Resalta
la forma irregular del cometa y las emisiones de
gas y polvo desde su superficie.

Comet by Rosetta (ESA). Image of Comet
67P/Churyumov-Gerasimenko obtained by ESA's
space probe Rosetta in 2015. It highlights the
irregular shape of the comet, and the emissions
of gas and dust from its surface.

Plutón por New Horizons (NASA). Imagen del
"planeta enano" Plutón obtenida por la sonda
espacial New Horizons de la NASA, en su pasaje
cercano del 14 de Julio de 2015. Se destaca la
variedad geológica de la superficie.

Pluto by New Horizons (NASA). Image of "dwarf
planet" Pluto obtained by NASA's space probe
New Horizons, during the flyby on July 14, 2015.
Diverse geologic regions are identified on the
surface.

currently estimated. Just over a century ago, in 1908, the famous
Tunguska explosion caused by a stony asteroid 50 to 80 m across
marked the largest such event in human history. The object hitting
Russia, which might be expected every millennium, resulted in a
10- to 30-megaton explosion (roughly 1,000 times greater than the
Hiroshima bomb, but without the radiation).[12]

An object of 140 m (460 ft) would cause significant devastation
and possibly a tsunami in the case of an ocean impact. Statistics
suggest that an impactor that size should be expected every 10,000
years on average. For an asteroid of 1 km (0.62 mi.), the rate of impact
is computed to be one every few hundred thousand years, but a recent
study of the Taurid meteor stream (which contains objects up to 1 km)
indicates an impact hazard of thousands or millions of times greater
when Earth passes near it: close passes will occur in 2032 and 2036.

This stamp sheet from
Uruguay features along the
top an actual photograph
of the Chelyabinsk meteor.
It commemorates the
Asteroids Comets Meteors
Conference of 2017. From
left to right are stamps
depicting Ceres, a comet,
and the largest Kuiper Belt
Object, Pluto.

Of the 20,000+ NEOs known, 2,100 are in the potentially hazardous category. Of these only 150 are larger than 1 km, and none are projected to be on a collision course with Earth as far as our computing power is able to determine. Such small objects are highly subject to gravitational perturbations from other planets and Earth itself, so it is not possible to accurately predict their positions into the distant future as we can do with a fairly high degree of accuracy for planets themselves.

What is much less heralded is the role heavy meteorite impacts have played in human history. An impact in the area of Australia 780,000 years ago dispersed impact glass across 30 per cent of the Earth. What effect this had on early humans is unknown, but it was certainly not negligible. There is strong evidence that a collision known as the Younger Dryas Impact was caused when a fragmented comet or asteroid that hit Earth 12,800 years ago in both the Northern and Southern Hemispheres;[13] a 2020 study found a site in Syria that is the first to document the direct effects of a cosmic impact at this time. It was subjected to temperatures in excess of 2,200°C (3,990°F) that wiped out a human settlement.[14] A study in 2017 has interpreted symbols in a temple at Göbekli Tepe, Turkey, as recording the impact. The carvings, on the Vulture Stone, date to 9000 BCE, and indicate a major impact from the sky in 10950 BCE. The image of a headless man is interpreted as a symbol for human disaster and extensive loss of life.[15] The recent discovery of the Hiawatha crater in Greenland, dating to this time, is now being employed to support the hypothesis of an extreme global event. Even though it is only 25 on the list of the largest craters, in the last 5 million years it is either the largest or second largest. What does seem certain is that something caused climate change and a significant extinction of large animals (such as sabretooth cats, mammoths and elephant-like creatures) at the time. The disappearance of human artefacts in South America coincides with this loss of big game, indicating that culture became extinct. In

a layer of sediment dating to 12,800 years ago from the Northern Hemisphere, Venezuela, Chile and Antarctica, scientists have found spherules containing platinum, iridium, osmium and gold, the very rare Earth metals that are abundant in some asteroids. Also found is a huge burning of biomass, the largest seen in a record that spans thousands of years. The Northern Hemisphere experienced cooling and expanding sea ice when one large fragment hit Greenland, while the Southern Hemisphere experienced a dry, warming phase as that area was hit as well. Research released in 2019 shows that the event was experienced in South Africa, where evidence of excessive platinum was found by Francis Thackeray and colleagues. 'We seriously need to explore the view that an asteroid impact somewhere on Earth may have caused climate change on a global scale,' Thackeray stated.[16]

In 2018 I visited a UNESCO World Heritage site, Ġgantija, dating to 3600 BCE. It is the grandest (most gigantic, as the name

Image of the Younger Dryas event dated to 10950 BCE showing people reacting to an airburst from a falling asteroid or comet. Art by Jennifer Rice, CometResearchGroup.org.

Ġgantija, dating to 3600 BCE, on the Mediterranean island of Gozo. Its inhabitants likely fled owing to the effects of an asteroid impact.

implies) of the structures built by the so-called Temple People on the island of Gozo, the second largest island of the nation of Malta. Their culture flourished for more than a thousand years, until it ended close to 2300 BCE. Why was this and other large temples on Gozo abandoned? A changing climate has been put forward as the most likely cause of this decline, but it is the proximate cause of this change that is of interest here. Some researchers believe a rain of debris from outer space was to blame. Like the event 12,800 years ago, a great comet or asteroid that underwent multiple disintegrations has been proposed. Dendochronology, the study of tree rings, has confirmed that trees across Europe were placed under tremendous stress in 2354 BCE, and it was over the next few decades that people migrated across Europe in search of more viable places to grow crops. Simon Stoddart, an archaeologist at the University of Cambridge, says, 'Malta is a living ancient story of what might happen in the present world.'

Roughly 650 years after this blast, another impact is thought to have devastated a civilization in Jordan, in an area known as Middle Ghor. Temperatures as hot as the surface of the Sun fried the area, as evidenced by zircon crystals (formed in less than a second), which have been found in pottery from the site. This finding, revealed in late 2018, shows that a civilization that had occupied this area for 2,500 years was wiped out in a moment. The area remained uninhabited for more than six hundred years after the destruction.[17] Could it happen to us now, in the twenty-first century?

Defending the Earth against an asteroid impact is not just in the realm of science fiction. Perhaps the strangest aspect of this threat involves the United States nuclear arsenal, and its need for lithium-6. The warheads of a nuclear bomb are multi-stage, and it is solid lithium-6 deuteride that serves as the fusion fuel for the second stage. Lithium-6 is also used to create the tritium gas that boosts the fission of plutonium in the weapon's primary stage. Given its crucial role, it came as a surprise to many when the National Nuclear Security Administration (NNSA) admitted in February 2018 that the country had lost its ability to produce lithium.[18]

Until a new factory is completed in 2026, the United States must rely on getting its lithium from recycled material from dismantled warheads. But the NNSA has exempted one very special warhead from recycling. Not to be tampered with are the secondaries of the five-megaton warhead from a missile dubbed Spartan. What is its intended use? None other than to defend Earth from an asteroid impact.

As the planetary scientist John Lewis wrote, 'Those asteroids that are most accessible to Earth for economic exploitation are usually the very same asteroids that have the highest probability of collision with Earth.' Trying to explode an asteroid to avoid a collision with Earth sounds like a good idea, but a study released in early 2019 indicates that asteroids are stronger than previously thought and would require more energy to shatter them completely.

Otherwise, they would most likely be able to reconstitute themselves as gravity pulls the fragments towards one another. While we have density estimates for many asteroids, their internal structure is unknown, and that could be a game-changer, turning hoped-for salvation into nothing but a vain attempt to save the planet.

The potential of asteroid impacts is already the subject of war games. A 20-metre (65 ft) asteroid passed 50,000 km (31,000 mi.) from Earth on 12 October 2017. It had been discovered in 2012, but for the purposes of a simulated potential hit on our planet, astronomers in 2017 assumed it had not yet been found. When it was 'found' with the Very Large Telescope in Chile in the summer of 2017, it was the faintest near-Earth object ever detected. The initiation of the war game began when a suite of telescopes that automatically survey the sky for moving objects (a survey system known as Pan-STARRS) detected the asteroid on 25 September 2017. After tracking the asteroid, the astronomers performed a continually updated risk assessment based on its size and composition. Mother Nature intervened to throw a spanner into the works when a fallen tree prevented the NASA Infrared Telescope Facility in Hawaii from observing the asteroid; and at the southeastern portion of American territory, the Arecibo radio telescope in Puerto Rico had been knocked out by Hurricane Maria. Overall the game was a success and showed that early detection of an Earth-crossing asteroid might give planners at least a few weeks to prepare for an impact.

The response of scientists and engineers to another fictitious asteroid with a 100 per cent chance of hitting Earth was studied in May at the 2019 Planetary Defense Conference near Washington, DC. In the war game, NASA and the space agencies of other countries decided to send six kinetic impactors to the asteroid in August 2024. While the main body was deflected, a fragment broke off and destroyed New York City. It is back to the drawing board for the next war games, scheduled for Vienna in 2021. Lembit Öpik, former Member of Parliament and grandson of the astronomer

Ernst Öpik, put the stakes in stark terms. In remarks made in late 2019, Öpik said, 'We do have to protect the Earth and that's because the chances of an impact are large enough to wipe out the human race and most other forms of life. It's essentially 100 per cent.'

To aid in deciding what option to employ in a deflection mission, researchers at the Massachusetts Institute of Technology unveiled a 'decision map' in 2020. Simulations show that with five or more years advance warning of an impact, the best course is to send two scout spacecraft to gather more data, followed by a projectile fired from Earth. Between two and five years, the most likely successful option is a single scout followed by a projectile. At one year or less, no option has a likelihood of stopping it.[19]

So what are the prospects of actually predicting where an asteroid might hit? Four objects, including 2019 MO, which hit Earth just south of Puerto Rico on 22 June 2019, have been spotted

The telescope Pan Starrs2, commissioned in 2014, sits atop the Hawaiian island of Maui at an altitude of 3,050 m (10,000 ft). It is used to survey the sky for asteroids.

in space before impact. A 3- to 5-metre Apollo object known as 2018 LA was discovered just 8.5 hours before it collided with Earth on 2 June. Fast work at the Center for Near-Earth Object Studies at NASA's Jet Propulsion Laboratory calculated it was likely going to hit southern Africa. Hours later a fireball travelling at 17 km per second was recorded by a security camera in Botswana, which led to the recovery of fragments five days later. A decade earlier, 2008 TC$_3$ became the first asteroid tracked in space before it entered our atmosphere. Like 2018 LA it was about 4 m (13 ft) in size and it also hit Sudan, leading to the collection of six hundred meteorites. Tracking a meteor back to its parent asteroid is an even more difficult task, but this was accomplished in 2020 when a binary near-Earth asteroid numbered 164121 was identified as the origin of a fireball that was seen over Kyoto, Japan, on 28 April 2017. Like the asteroid Gault, it must be shedding small particles that appear in our skies as meteors.[20]

Of all the ways asteroids might be useful to humanity, none is more important than the future of humanity itself. The United States has established a Space Force, which may encompass all the space operations currently divided between the various branches of the military. In a 2018 CNN interview, astronomy popularizer Neil deGrasse Tyson weighed in to emphasize its potential importance.

> There are future needs and concerns about what a Space Force would have to think about, such as asteroid defence. That's a complete space-based threat to not only the nation but the world. It may be that the co-operation of the world in space is to save civilization and that could be the greatest force of peace there ever was.

So instead of ending civilization, which an asteroid impact has the potential of doing, averting such an impending catastrophe may prove to be the catalyst for a new world order.

Asteroid Surveys

In 2005 the U.S. Congress mandated NASA to track 90 per cent of near-Earth objects larger than 140 m. These are known as Potentially Hazardous Asteroids (PHAs). In 2020 it was estimated that we were only a third of the way to achieving that goal. With current efforts the goal will not be reached for several decades, which has led to the development of a space-based telescope, the NEO Surveyor, to search for NEOs. Its prime mission is to find the mandated 90 per cent within ten years, gathering both physical and orbital data on them. However, NEO Surveyor has consistently been denied full funding. In late 2018 former NASA astronaut Rusty Schweickart of the Apollo 9 mission publicly called for the full funding of NEOCam, which received only $35 million in 2018 – enough to keep the project alive, but not enough to launch. 'It's a critical discovery telescope to protect life on Earth,' he said. NASA announced in September 2019 that it will proceed with NEO Surveyor, establishing a launch date as early as 2025. Thus we won't have the 90 per cent tracking data until at least 2035.

Meanwhile a ground-based effort known as the Catalina Sky Survey is very productive. With two telescopes in Arizona, it began making asteroid discoveries in 1999 and currently accounts for nearly half of the NEOs found annually. It is credited with discovering more than 50,000 numbered asteroids, including 10,000 NEOs. Pan-STARRS currently operates two telescopes in Hawaii; as a NASA-funded project it concentrates on the detection of NEOs. 'Oumuamua has been its most important discovery. Another Hawaii-based project, with telescopes on Mauna Kea and the Big Island, is ATLAS. As of late 2020 it will offer true pole-to-pole coverage every 24 hours; 'its ability to look at awkward places in the sky', said Larry Denneau of the ATLAS project, may find asteroids that are missed by other surveys.

Three earlier efforts deserve mention. LINEAR, a project of MIT's Lincoln Laboratory, NASA and the United States Air Force, is

credited with 149,793 numbered asteroid discoveries from 1998 to
2011. NEAT, which ran from 1995 to 2007, found 41,359 asteroids.
It was a joint project of the Jet Propulsion Laboratory and NASA.
Finally, LONEOS observations ran from 1998 to 2008, under the
direction of Ted Bowell. Based at Lowell Observatory, with NASA
funding, it found 22,511 asteroids.

In 2009 NASA launched a spacecraft dubbed WISE (Wide-field
Infrared Survey Explorer). Once its primary mission to survey distant
infrared sources was completed in 2013 it was repurposed as a
detector of near-Earth objects and comets. 'It was never designed to
look for asteroids,' explains Amy Mainzer of the NEOWISE science
team, 'but it is sensitive to warm NEOs, which are easily visible as
orange dots in a background of blue stars.' NEOWISE observed
32,991 different objects during the first five years of its survey. A
big advantage of observing asteroids in the infrared is that 'we can
see the heat these objects are emitting and can determine their size
to within 10–20 per cent.' Over the past few years there has been a

This artist's concept shows
the Wide-field Infrared
Survey Explorer, dubbed
NEOWISE, in its current role
as an infrared detector of
near-Earth objects. The NASA
spacecraft was launched in
2005.

major question about just how many large asteroids are out there. The work of NEOWISE has confirmed that we do know the existence of 'nearly all those 1 m and larger objects. This is good news for Earth, bad if you make disaster films,' said Mainzer.

Finding asteroids is one thing, but tracking them to determine their orbits, the branch of astronomy known as astrometry, is another. Finding prediscovery images is a big help when calculating an accurate orbit, but the most densely concentrated asteroids in the inner main belt, known as Hungarias, often mimic the movement of NEOS. This prompted prediscovery expert Richard Wainscoat to lament, 'Hungarias are the curse of the universe!'

The brightest sets of dots in this 13 December 2019 image by NEOWISE belong to asteroids 97 Klotho and 468 Lina. Both orbit in the main asteroid belt between Mars and Jupiter, while smaller, more distant asteroids can also be seen passing through the image.

Based at the University of Arizona, Robert McMillan as Director of Spacewatch described to me the most venerable asteroid survey project. McMillan said Spacewatch got started in 1992,

when Tom Gehrels revived his 1970s wish to build a 72-inch telescope dedicated to observations of asteroids. It took ten years of hard tribulations to get the money and build it, but we finished it and began observing with it in 2002 (NASA has supported the project since its inception). Having such a 'large' aperture, by the limited standards of the asteroid community, reinforced the already-recognized niche of Spacewatch as the group that could observe to fainter magnitudes than the surveys that were operational in the late 1990s and early 2000s. Indeed, when the three wide-area surveys of LINEAR, NEAT and LONEOS began finding NEOs rapidly in 1998, we (Spacewatch) decided to play our card of faint detection by redirecting our

Tom Gehrels, in the late 1980s, with the 36-inch (0.9 m) Spacewatch telescope in Arizona.

efforts to follow-up astrometry. NEOs tend to get discovered during their close approaches to Earth. Consequently, to get enough observations over a sufficient arc to get accurate orbital elements, they must be followed as they recede into the distance and become fainter. So the telescopes doing follow-up should be larger than the ones used to make the original discoveries.

Astrometric follow-up thus defined the future of Spacewatch from 1998 to the present and beyond. In 2002 we were recruited by the WISE spacecraft mission team as the group best equipped to do visible-light astrometry on the NEOs that WISE would be discovering. Being associated with a spacecraft mission was our entry card to getting time on competitively allocated telescopes from 2010 on. Since then we have been awarded more than 200 nights on telescopes of 2.3-metre to 10-metre aperture. Spacewatch has made three-quarters of all the observations of PHAs while the objects were fainter than visual magnitude 22.5. If our success in getting time on large telescopes is any indication of the future of Spacewatch, then we will be well situated to continue our usefulness into the era of the Vera Rubin Observatory and NEOCam.

The Vera Rubin Observatory (VRO) will house a telescope now being built in Chile. This telescope, with its wide-field CCD camera containing 3 billion pixels, will be able to survey the entire sky in three nights. 'Observations will be 30 seconds long, and 30 seconds later the system will issue an alert if anything has moved,' said Juric Mario of the VRO project. With operations set to commence in 2022, it is expected to increase by a factor of ten the known NEOs that are larger than 30 m (100 ft), and within a decade of operation it will image 250,000 asteroids as small as 15 m (50 ft). Mainzer said, 'If you want to find all the 140-metre-sized objects, both the VRO and NEO Surveyor together will do it.'

Even the world's largest single-aperture telescope has been employed to study asteroids. In 2019 the Gran Telescopio Canarias (10.4 m mirror) made its first observation of an NEO (2019 DSI) as it joined the Planetary Defense programme of the European Space Agency (ESA). The NEO was discovered by Catalina on 28 February 2019; data from the giant telescope shows it will pass 165,000 km (102,500 mi.) from Earth on 26 February 2082. ESA, in collaboration with the Italian Space Agency, will locate the first Flyeye telescope on Monte Mufara in Sicily in 2020. Even though it is officially called NEOSTEL, for Near-Earth Object Survey Telescope, it is dubbed Flyeye because it mimics the eye of a fly in that it will

The dome of the Vera Rubin Observatory (VRO), pictured under construction at sunset in May 2019 in Cerro Pachón, Chile.

employ multiple cameras and optics, splitting the sky into sixteen smaller images in order to expand its field of view. The goal of the proposed network of four Flyeye telescopes will be to automatically scan for NEOs. Southern sky coverage for the ESA programme will be provided by the Zadko telescope in Western Australia, which joined the programme in December 2019. ESA's Gaia satellite (launched in 2013) also has the potential to find more asteroids (its first discovery was announced in May 2019), but primarily it is only being used to improve photometry and spectral data on existing objects. After scientists sifted through four years of data collected by the Dark Energy Survey camera, mounted on a telescope in Chile, it was announced in 2019 that 139 new TNOs had been discovered. Eight of these have a semi-major axis greater than 150 AU. Even though the camera is designed to gather data on the cosmological

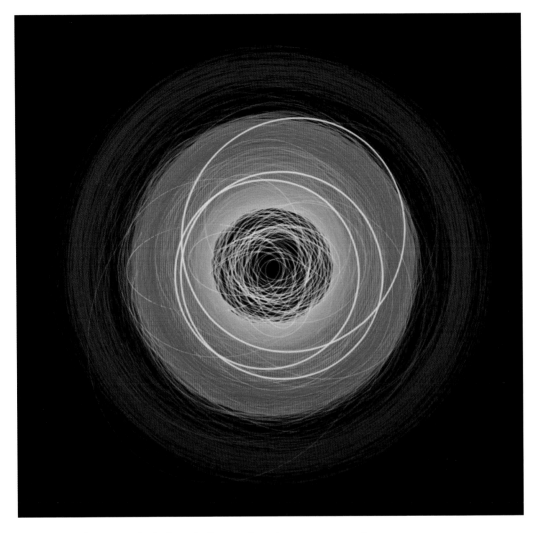

study of dark energy, its infrared observations have proven to be an unexpected treasure trove for distant solar system objects. 'In recent years', Don Davis explained to me, 'large-scale surveys have discovered hundreds of thousands of asteroids and soon the total will exceed one million. New observational techniques and additional spacecraft missions will ensure that the excitement of recent decades will be propelled into the future.'

The Gaia satellite, whose main mission is to chart stars in our galaxy, also began making asteroid discoveries in 2018. The orbits of the first three of these are shown in white, while the red orbits represent the entire main belt population.

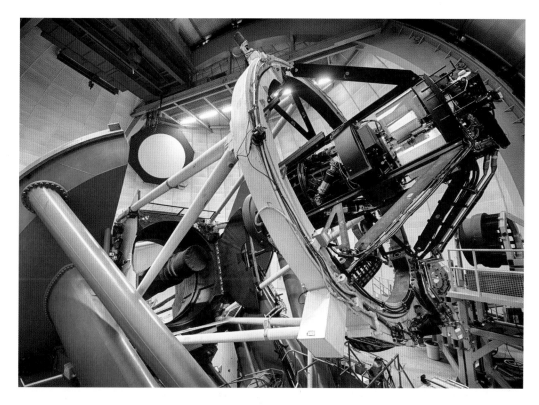

The Dark Energy Survey (DES) camera attached to the 4-metre Blanco Telescope in Chile has discovered TNOs.

Meteorites and Life on Earth

David Fisher of the University of Miami dates the birth of modern planetary science to 1942, when William J. Arrol, R. B. Jacobi and Friedrich Paneth made a startling discovery in several iron meteorites. Each had more helium-4 than expected. Since this helium originated from nuclear reactions induced by cosmic radiation, it was an indicator of how long the meteorites had been exposed to that radiation. Geological evidence at the time suggested the Earth was at least 2 billion years old, but the meteorite evidence yielded an age of 4.5 billion years, indicating the universe was older than anyone had previously believed. Once it became obvious that research on meteorites could probe the origins of the universe, and also the solar system, they became a serious object of study.

This is directly relevant, as meteorites largely come from asteroids. Long before the Japanese spacecraft Hayabusa returned to Earth a tiny amount of asteroid 25143 Itokawa in June 2010, planetary scientists were able to study samples of asteroids in the form of meteorites. Of course, these are not pristine samples, as the chunks of rock or metal were heated during their passage through our atmosphere. Several papers in the life sciences during the last few years have linked the formation of DNA (and thus the beginnings of life on Earth) with the energy delivered by asteroid impacts during the LHB 4 billion years ago. In light of the fact that there was no LHB, claims tying the origin of life to asteroids must be dismissed. It is important for the lay reader to know this and not be misled, as television documentaries and other media reports will no doubt base their assertions on the reality of the LHB for many years to come.

This does not mean that asteroid impacts on terrestrial planets have no place in the discussion on the origin of life. The habitability of Earth is strongly tied to its very late stage of growth, based on

A study of the peanut-shaped asteroid Itokawa reveals two sections of it have different densities. It may have formed from two asteroids that bumped together and merged.

1750 kg/m³

2850 kg/m³

research conducted in 2019 that has enabled scientists to date the delivery of carbonaceous chondrite material that is now trapped in Earth's mantle.[21] Research released in January 2020 by Bettina Schaefer and colleagues indicates that bacterial life in particular thrived in the impact crater that wiped out the dinosaurs;[22] and in a 2019 study, Rafael Navarro-Gonzalez and colleagues found that asteroids may have produced the key ingredients for life on Mars. After simulating the early Martian atmosphere, 'the big surprise', he stated, 'was that the yield of nitrate increased when hydrogen was included in the laser-shocked experiments that simulated asteroid impacts'.[23] In 2015 the Curiosity rover on Mars discovered nitrate (NO_3) in the Gale crater, but until now its origin had remained a mystery. Why are nitrates important? Because they are fixed forms of nitrogen, a necessary gas for the development of life. Combined with evidence that Mars once had flowing water on the surface, the presence of asteroid-induced nitrogen brings us one step closer to understanding the possible development of life on Mars as early as 4.2 billion years ago and brings us closer to understanding development of life on exoplanets. The 14.8-million-year-old Ries crater in Germany is being studied as an analog to what happened on Mars as the rocks in Ries have revealed several nitrogen isotopes created by an asteroid impact. The Perseverance rover is slated to land in the Jezero crater on Mars in 2021 to search for similar nitrogen isotopes in an effort to establish the conditions for ancient life there.[24] Mars is still being hit by asteroids: an image taken in 2009 shows no craters in a certain area of Noachis Terra, while a 2016 image by Mars Reconnaissance Orbiter does. The impact reveals the lighter-toned materials that lie beneath the surface. The fact that the thin Martian atmosphere was able to break up the small asteroid suggests it was comprised of weak materials such as stone, rather than metal.

In late 2018 scientists at NASA's Ames Research Center in California reported on laboratory experiments which suggest that

certain sugar molecules required to form life may have originated from asteroids. A direct impact by an asteroid, or perhaps smaller fragments in the form of meteors, could have delivered these ingredients to our planet in its early years. While only one sugar has been identified in meteorites, the study indicates that several sugars necessary to the development of life form naturally in meteorite parent bodies – asteroids and comets.[25]

The idea that Earth received its life-giving water from ancient impacts received a boost in 2020 when planetologists at Münster University released a study of the small Flensburg meteorite that fell in Germany on 12 September 2019. Flensburg contains minerals that required liquid water to develop, proving that planetesimals contained water 4.56 billion years ago.

If Earth is ever hit by a major extinction event, many fragments blasted off our planet will carry biota with them, so when they eventually return to the surface they may reseed the planet. The cycle of existence on Earth and possibly Mars, from the origin of

Craters surrounded by white ejecta blankets mark a recent impact site on Mars. The image depicts 21 craters ranging from 1 to 7 m (3 to 23 ft) in diameter, distributed over an area that spans 305 m (1,000 ft).

Murchison meteorite at the National Museum of Natural History (Washington, DC), which contains pre-solar grains measuring from 2 to 30 micrometres that are the oldest material found on Earth. The Natural History Museum (London) also has a Murchison fragment.

life to the extinction of life, may thus be mediated by asteroids, making a thorough study of them of the utmost importance for biologists, chemists, physicists and astronomers. The value of studying meteorites was emphasized in 2020 with the first dating of interstellar dust. It was found that grains from the 1969 Murchison meteorite, which fell in Australia, were as old as 7 billion years, before the start of the solar system 4.6 billion years ago.[26] Both the origins of life and the solar system itself will be major topics of research in asteroid studies in the twenty-first century.

Meteorites and Impact Craters

Asteroids have dominated the impact history of Mercury and Mars. Of the 61,000 meteorites that have been found on Earth, 224 were blasted off Mars by asteroid impacts. One of those, dubbed Black Beauty, is 2 billion years old and is now being studied to further our understanding of ancient Mars. The process also worked in reverse: in 2019 an analysis of Moon rocks brought back by the Apollo 14 mission (which I saw launched from Cape Kennedy in 1970) revealed that a 2 g fragment from the lunar surface actually originated 20 km (12 mi.) below the Earth's surface. The fragment, composed of quartz, feldspar and zircon (most unusual on the Moon), was hurled into space by a huge asteroid impact 4 billion years ago, eventually landing on the Moon. The Manicouagan crater in Canada, which is defined by an annular lake, was created by an asteroid impact 214 million years ago, but much earlier evidence for impacts on our own planet is hard to find: the 13-kilometre-wide

(8 mi.) impact crater of a 1-kilometre asteroid strike in Scotland, which happened 1.2 billion years ago, was only identified in 2019. The impact happened between what is now the mainland of Scotland and the Outer Hebrides.[27] Around 600 to 700 million years ago the Earth lost most of its craters, probably owing to a massive global-scale erosion event, but such ancient craters, called crayons, exist on stable continental cores.[28] In November 2018 an impact crater was for the first time discovered under ice. This 30.5-kilometre-wide (19 mi.) crater is just under a kilometre beneath the Greenland ice sheet, specifically the Hiawatha glacier in northwest Greenland. It is estimated the impactor was 0.96 km (0.75 mi.) across. In February 2019 an even larger crater, measuring 35.4 km (22 mi.) across, just 183 km (114 mi.) from the Hiawatha crater, was discovered. It is unlikely they hit at the same time, as the ice over the Hiawatha crater is no older than 12,800 years, while that over the larger crater is at least 79,000 years old. The oldest recognized meteorite impact structure is the 70-kilometre-wide (43 mi.) Yarrabubba crater in Western Australia, dated at 2.229 billion years. Timmons Erickson, leader of a team who dated the structure, in research published in 2020, explained: 'there is no immediate evidence for an impact crater. This is due to the antiquity

Slice of an iron meteorite, which came from an asteroid like 16 Psyche. This one was found in 1890 in New Mexico, USA. The pattern shown here cannot form in rocks on Earth. It is known as a Windmanstaetten Pattern, named after Count Alois von Beckh Windmanstaetten, who discovered it in 1808 while treating a meteorite with a solution of nitric acid. This specimen is in the Vatican collection, housed at Castel Gandolfo.

Canada's Manicouagan crater, 100 km (62 mi.) across, was created by an asteroid impact. It is shown in this image taken on the space shuttle mission STS-9, November–December 1983.

of Yarrabubba; it is believed that at least 4 km of erosion have occurred since the impact. Shocked minerals and a geophysical anomaly are all that reveal the original impact structure today.'[29]

This highlights the difficulty in finding evidence for ancient craters on Earth's surface. Research in 2018 indicated that much if not all of our planet was covered in ice 600 to 700 million years ago. This massive ice blanket dumped a third of the entire crust of the planet into the oceans, erasing much of the evidence for craters.

An understanding of a link between meteorites and asteroids goes back to the earliest years of the study of asteroids. In 1805 Johann Christian Wildt confidently asserted that the stones which have fallen on Earth are 'the ruins of some globe which has been destroyed, and which revolve round the sun till, sooner or later, they fall in with a planet. They without doubt belong to the group of Ceres, Pallas, &, and thus we see how it is that their appearance and composition bear such general resemblance to each other.' Wildt was a professor of philosophy at Göttingen, the same university where Gauss worked. But Wildt's theory, which had the merit of correctly attributing at least some meteorites to the asteroids, found little favour in the first half of the nineteenth century. Olbers, whose opinion was highly regarded across Europe, distanced himself from Wildt's meteorite origin theory in a paper of 1837. By 1849 the tide had begun to turn: the great Prussian naturalist Alexander Baron von Humboldt called meteorites 'the smallest of all asteroids', and correctly said 'they may come from some crumbling asteroid'.[30]

In 1881 a lawyer, Otto Hahn, published a book in support of the theory, advanced by several people including Sir William Thompson (later Lord Kelvin) in 1871, that life probably originated on Earth from seeds brought to it on meteorites from the ruins of another world. Hahn claimed that meteorites were full of fossil debris, with some fossils of earthly origin and others transported from different planets, but Stanislas Meunier of the French Academy of

Sciences demonstrated that all the supposed fossil debris was merely clustering crystals of eulytite (monoclinic bismuth silicate). Meunier had his own theory for the origin of meteorites: he thought they were 'the shattered morsels of a satellite smaller than and perhaps subordinated to our Moon', in the words of Nevil Story Maskelyne explaining the views of Meunier to an English audience.[31]

Here we make the acquaintance of the grandson of Nevil Maskelyne, who was one of the first people in England to see Ceres and Pallas; Nevil was a close friend of both William Herschel and his sister Carolyn. Nevil had a daughter, Margaret, whose son Nevil Story Maskelyne (the family adopted the surname Maskelyne after the younger Nevil came of age) became a noted geologist with a great interest in meteorites. In 1870 he dismissed the conclusion by Meunier that iron meteorites fell only in ancient human history, and that we now see only stony meteorites hitting Earth. Nevil Story wrote in that year of a future time when 'we shall be able better to decipher the characters in which the history of the meteorites is written, a history which assuredly is engraved, though in hieroglyphic language, on these messengers from space to our world.' Despite the florid Victorian prose, he was not far wrong. A scientific paper of 1978, describing a meteorite from Brazil, states that 'large, hieroglyphic schreibersites dominate the structure.' These schreibersites are a mineral common in iron nickel meteorites. This is not the place to explore the study of meteorites, but the century between 1870 and 1970 placed their study on a firm scientific basis. We now know that the origin of meteorites is linked to both asteroids and comets (a meteorite from an asteroid, discovered in Antarctica in 2002, has even been found to contain a fragment of an ancient comet), while a few of them were blasted off the Moon and Mars when asteroids struck those bodies millions of years ago. All the meteorites found so far originated from between 95 and 148 primary parent bodies.[32] So when you look at meteorites in a museum, you are likely to be looking at bits of an asteroid.

Comets

In 1824 Olbers wrote about the possible connection between comets and asteroids.

> In terms of form, shape and nature of comets of short sidereal periods, there seems to be no difference with those of very long sidereal periods. They have absolutely no similarity in external appearance with the four newly discovered planets. But whether the comets of very short periods – Encke's, for instance – are not related to the asteroids in some way or another, ought not be the subject of any conjecture until these comets have been seen more often returning to their perihelion, so that one may begin to learn about the possible changes which they suffer over time due to their physical composition.[33]

It was not until a century later, in 1932, that Ernst Öpik (who later worked at the Armagh Observatory in Northern Ireland) proposed that comets originate in a cloud orbiting in the distant solar system. This region, now known as the Öpik-Oort cloud to recognize the same suggestion made in 1950 by the Dutch astronomer Jan Oort, covers a vast range from 2000 AU to 200,000 AU. This is the home of countless objects termed 'planetesimals', a mixture of comets and asteroids that has persisted for billions of years. Some of these objects can be sent into the inner solar system by forces exerted by passing stars in our Milky Way galaxy, but while in the cloud they are too faint and distant to be studied with current technology. In 2019 a small particle in a carbonaceous meteorite was identified as a cometary building block, likely originating from the early stages of planet formation.[34]

Stars come in different colours and so do asteroids. The hottest stars shine with a blue light. Have you ever heard of a blue asteroid? Well, there are a few, and one of them provides our best link between asteroids and comets. A study released in late 2018 of 3200 Phaethon shows that it is not only the bluest asteroid known, but the source of the annual Geminid meteor shower. Cometary sources of meteor streams are typically red, and asteroids are typically dull grey to red, so clearly something strange is going on here. The surface of Phaethon is uniformly blue (it is designated as a B-type asteroid); the fact that its surface heats up to 800°C (1,500°F) every time it passes close to the Sun probably accounts for that. At these perihelion passages it releases a small dust tail like a comet, but otherwise it appears inert like an asteroid. Phaethon may be related to Pallas, which is the largest blue asteroid, but it is only half as reflective as Pallas. Japan is planning to send a spacecraft called DESTINY+ to Phaethon around the year 2025.

FOUR

ASTEROIDS IN POPULAR CULTURE

In 1806 an article appeared in *Messenger of Europe*, published in Moscow. It touted the fact that daily circulation of Moscow newspapers had reached 8,000, and the sale of books had also reached new heights as the Russian public was becoming more literate. The author tells us that romance novels were in the greatest demand. Far from decrying the fact that serious books were not so popular, he observes that 'modern romances abound in various kinds of knowledge. An author, to fill up several volumes, is obliged to have recourse to all methods and almost all sciences.' Then comes a most remarkable assertion. 'I am persuaded that in certain German novels, the new planet, *Piazzi*, will be more circumstantially described, than in the Petersburg newspaper.' In modern parlance, he is writing that the newly discovered Ceres would be more fully described in a mere novel in Germany than a newspaper in Russia since novels are full of up-to-date scientific knowledge.[1]

Asteroids entered the public consciousness in many ways. A satirical story appeared in an 1851 issue of *The Carpet-bag*, a Boston publication that poked fun at Southerners. The reader is told of a certain Zedekiah Bump, who had just returned home from a visit to Boston where he had attended an astronomy lecture. There he saw a mechanical device known as an orrery that shows the orbits of the planets around the Sun. The local science society urged him to give a talk on astronomy, so Zedekiah decided to build an orrery. Not having finely made machine parts, he used whatever he could find from his father's food store. A large pumpkin stood in for the Sun. 'Gradually Mars, Vesta, Astraea, Juno, Ceres and Pallas sprang into being, as an onion, a long-red, a golden pippin, a pumpkin-sweet'n, a Chenango, and a rootabaga.' Even though all five known asteroids made the cut, Zed decided not to include Saturn or Herschel (Uranus) as he could not figure out how to depict the Saturnian rings, or all the moons of those two planets.[2]

Many people in the nineteenth century, when literacy was on the rise, may have first read about asteroids in poetry. Dozens of poems celebrated asteroids. In 1811 Pastor Gottlob Schulze of Polenz, just east of Dresden, published a book in German entitled *Das Sonnensystem, so wie es jetzt bakannt ist* (The Solar System, as It Is Now Known), which incorporated a Latin poem he had published a year earlier. This is a fine example of how poems were designed to inform the public of scientific information about the asteroids. It alludes to Bode's Law, which predicted that one object would be found in the 'excessive space' where we now see four. It goes on to name the discoverers of the first four asteroids, the locations of the discoveries, the orbital periods of the asteroids and even the theory of their origin.

Among the ancients, they judged Jupiter to be solitary.
But when our generation should look upon excessive spaces,
And be withdrawn from the usual law of standing apart:
A suspect matter advises men to search deeper;
Prepared with talent, skill, equipment, they see the truth,
They see four before the one, amazing to say,
Which Olbers judges to be fragments of a shattered planet.
Thus the Sicilian Piazzi uncovered Ceres by chance;
She flies across her circuit in twice six times twenty weeks.
Thus did wise Olbers of Bremen add Pallas;
She wings about her orbit in the same time as Ceres.

Thus did lucid Juno reveal herself to Harding at Goettingen;
Her path is three and ten times five months.
Happily did Olbers catch sight of the looked-for Vesta;
Through forty-three months she takes her road.
The circuit of these four, than which nothing is more puzzling,
Would scarcely have been explained, unless you, Gauss,
Standing so much above us, would have demonstrated it, may
you never pass on.[3]

The most famous appearance of asteroids in an English novel
happened in 1914 when Arthur Conan Doyle wrote a Sherlock Holmes
mystery entitled *The Valley of Fear*. Of Holmes's perpetual nemesis,
Professor Moriarty, Sherlock Holmes says to Dr Watson, 'Is he not the
celebrated author of the *Dynamics of an Asteroid*, a book which ascends
to such rarefied heights of pure mathematics that it is said that there
was no man in the scientific press capable of criticizing it?'

This fictional book within a fictional book so inspired the great
author Isaac Asimov that he wrote a mystery story in 1976 based
on the *Dynamics* book. A character in Asimov's tale comes to the
conclusion that Moriarty wrote about the origin of the asteroids as
being an explosion of a planet. It is further reasoned that, true to
his nature as a diabolical mastermind, Moriarty planned to use his
calculations of the primordial planet explosion to destroy Earth.
Thus the title of Asimov's story: 'The Ultimate Crime'.

In 1823 the English writer Charles Bucke neatly encapsulated
why we all find asteroids so appealing.

When we meditate on the comparative diameters of Uranus,
Saturn, Jupiter, and the Sun, we are astonished; but our
curiosity is much more excited by the diminutive proportions
of the Asteroids. They best suit the limited compass of our
understanding. Man most admires the great; but he most loves
the little.[5]

The most famous book of the twentieth century to incorporate asteroids is *The Little Prince*. Since first appearing in print in 1943, the book by Antoine de Saint-Exupéry has sold more than 200 million copies. The book is not really about asteroids at all, as they make an appearance solely as a plot device. We learn that the eternal child, known as the Little Prince, was born on asteroid B-612. 'This asteroid has been sighted only once by telescope, in 1909 by a Turkish astronomer,' the fictitious tale reads, 'who had then made a formal demonstration of his discovery at an International Astronomical Congress. But no one had believed him on account of the way he was dressed.' The boy prince travels to various very tiny asteroids, visiting there a succession of characters (only one per asteroid, as they are not large enough for more) that advance the story. It is the ultimate children's book, with moral tales for adults worldwide.

Perhaps because children are small, they relate especially well to asteroids. The Jim Henson TV production *Dinosaur Train* is geared towards educating youngsters, with Canadian palaeontologist Scott D. Sampson as a scientific advisor. In a 2015 episode the Conductor of the train takes two dinosaur parents and their kids in a zeppelin to view a crater. 'Not all craters are barren and empty,' says the Conductor, 'some fill with water and become crater lakes and other craters have jungles growing in them.' One of the kids ponders out loud, 'What are the odds an asteroid would hit us?', the same question humans are asking now. Amy Mainzer is the scientist who appears in a CGI animated show aimed at kids, *Ready Jet Go!* A 2018 episode featuring an 'asteroid belt space race' concluded with her explanation of the difference between meteors and meteorites.

Science Fiction

The role of asteroids in pulp fiction magazines of the early twentieth century did not spring into existence in a literary vacuum.

Its parentage may be uncertain, but there is no doubt that its grandfather was Konstantin Tsiolkovsky.

His most extensive fictional work, *Outside the Earth* (*Vne zemli*), was published in the midst of the First World War in 1916. It was really the final effort in a trilogy. The first, a short story, 'On the Moon' ('Na lune'), was published in 1893; the second, a novella, *Dreams of the Earth and the Sky* (*Grezy o zemle I o nebe*), published in 1897, describes mankind's settlement of space, complete with characters who mine asteroids. In this, Tsiolkovsky postulated the existence of plant-like sentient beings on an asteroid. 'They are far ahead of us in science and perform amazing feats of planetary engineering, like dismantling asteroids into rings or "necklace" formations so that they can take advantage of very low gravity.'

Together, the trilogy embodies the genesis, evolution and final synthesis of the technological aspects of space flight. *Outside the Earth* deals with the final phase of departure from the Earth, with the resettlement of humanity in so-called 'hothouses' suspended in the cosmos. To sustain this colonization, the space pioneers realize that the best sources of raw materials are asteroids. It was in this tale of science fiction that humans of the twentieth century first ventured into the asteroid belt, and there was no turning back. Fascination with asteroids proved to be an enduring motif that caught the imagination of many American writers of the next generation who were published in pulp science fiction magazines.

Asteroids as a plot device blossomed in these stories, making asteroids an integral part of American pop culture. During the last twenty years I have collected a hundred pulp fiction magazines from 1929 to 1959 in which asteroid-related stories were published. Many of the elements that recur in the early pulp stories of the 1930s can be found in Tsiolkovsky's writing about the riches and mysteries of the asteroids. These elements include vast mineral wealth, their problematic origin and what that might mean for the future of the Earth, and the human colonization of space.

The pulp fiction magazine *Wonder Stories*, September 1931 issue, depicts 'An Adventure on Eros'.

Mineral wealth was just one of many tropes that ran through the asteroid stories. Adventure, geared to the teenage male reader, was much in evidence. Imagine gazing at the exciting cover of the September 1931 issue of Wonder Stories in the midst of the Great Depression, when life was bleak for many citizens of the United States. Written by J. Harvey Haggard, the main story is entitled 'An Adventure on Eros', and the cover image shows a spacecraft leaving that asteroid, apparently fleeing a horde of metal men that live on Eros. Not content with giving young readers a thrilling adventure, a note from the editor, Hugo Gernsback, explains that 'the greatest impressions on our minds are made by those experiences that we suffer, that often mean life or death to us . . . Perhaps, as this ingenious story points out, the process proposed by Mr Haggard, could be carried further, and education by "experience" become an established method of teaching.'

Asteroids are also popular fodder for games. When Atari's arcade game *Asteroids* was released in 1979 it quickly became an American cultural icon. The quintessential shoot-'em-up game let a whole generation learn it was fun to blast asteroids. That same year a board game, *Belter: Mining the Asteroids, 2076*, envisioned a new Wild West in the sky. '150,000,000 miles from Earth. One of the last frontiers in the Solar System is certain to be the Asteroid Belt. By the time that regular access to the region is possible, the advance of technology can conceivably make individuals capable of conquering it. Such a frontier would necessarily be lawless and bountiful . . . the ideal domain of the classic frontiersman, now called the Belter.' The board game *Asteroid Pirates* was based on 'ship to ship combat in the asteroid belts', while a game simply titled *Asteroid* envisioned going 'into an asteroid station, against a deranged computer, to save a threatened world'.

The most visible way asteroids have entered (some might even say infiltrated) pop culture is through films and television. In 2017 a TV series entitled *Salvation* was aired, produced by Alex Kurtzman, who has also brought us the new *Star Trek* show *Discovery*, which includes mention of an asteroid made of dark matter. The main premise behind *Salvation* is that a large asteroid is on a collision course with Earth. 'It's amazing what people can accomplish when an asteroid is hurtling toward them,' says U.S. Vice President Darius Tanz in a 2018 second season episode, 'Let the Chips Fall'. 'Behold the saviour of the human race,' he declares when looking at a 'rail gun designed to propel one-ton slugs of depleted uranium moving at 30 times the speed of sound into the asteroid'. This is the kinetic-impactor scenario, trying to nudge the asteroid so it sails past the Earth instead of hitting us. (In real life, however, modifying an asteroid orbit would need to be done years in advance. The first actual attempt to kinetically alter an asteroid's orbit is slated for 2022, as explained in Chapter Five.) The TV show *Expanse* premiered in 2015. The author of the fantasy series *A Song of Ice and Fire* (adapted for television as *Game of Thrones*), George R. R. Martin, wrote in 2018 that *Expanse* 'was the best space show on television. For hard-core S.F. geeks like me, there's nothing quite like a space show. Weird worlds, alien proto-molecules, gritty asteroid miners,

Asteroid Commander. From 1953 (pre-Space Age) comes a syrup bottle that doubles as a coin bank. It is one of a set of twelve made by the Space Foods Co. of Baltimore. Each bottle, with names such as Orbit Admiral, Space Navigator and Space Commander, is shaped like a spaceman wearing a helmet.

The 1979 board game *Belter: Mining the Asteroids, 2076*, by Frank Chadwick at the Game Designers Workshop.

From 1981, *Asteroid Pirates* features a depiction of combat in the asteroid belt by John Hagen.

The 1980 board game *Asteroid*, futuristically set in the year 2007. It was designed by Frank Chadwick and Marc Miller and released by the Game Designers Workshop.

cool spaceships.' There is even a comedy that premiered in 2020, *Don't Look Up*, about two astronomers who try to warn the world about the dangers of an approaching asteroid.

Films have been depicting collisional events for many years. Examples such as *The Day the Sky Exploded* (1958) and *Meteor* (1979) starring Sean Connery are considered classics. The former is considered to be the first Italian science fiction film. The trailer for the Connery movie unscientifically employs sound effects to emphasize the motion of an asteroid through space, while the narrator says, 'Its power is greater than all the hydrogen bombs. Its speed is higher than any rocket ever conceived. Its force can shatter continents. Its mass can level mountain ranges. It cannot think.

It cannot reason. It cannot change its course!' In both films some asteroid fragments wreak havoc on Earth, while the threat of mass extinction is eliminated with the use of nuclear missiles.

Two films released in 1998 dealt with impacts. *Deep Impact* envisaged a comet on a collision course with Earth, while *Armageddon* had an asteroid the size of Texas ready to wipe out humanity in eighteen days. The solution in *Armageddon* was to divert the asteroid by exploding a nuclear bomb, deposited deep inside. But to disrupt an asteroid that size would require the same amount of energy as is produced by the Sun. While scientifically inaccurate, it had great box-office appeal.

Even the world of anime is in on the act. In 2018 asteroid 113405 was named Itomori after a fictional Japanese town that was destroyed by the impact of a comet fragment in the film *Your Name*. (Being killed by a comet fragment instead of an asteroid fragment is a distinction without a difference. Some 6 per cent of near-Earth asteroids are extinct comet nuclei and the origin of many meteor streams from comets shows they contain more rocky material than the simple dirty snowball hypothesis would lead one to believe.) Another avenue of popular expression is modern art: a 2009 work entitled *Epic 1* by Teresita Fernández was inspired by a meteor shower, and the popularity of asteroids has led America's Jet Propulsion Laboratory, where many real unmanned space missions are controlled, to release a travel poster touting Ceres as the 'Gateway to the Outer Solar System'.

The first time a collision was ever predicted between two named objects in the solar system happened in 1992 when Jupiter was hit by fragments of a comet that was discovered by Carolyn Shoemaker. In 2019 Dagomar Degroot at Georgetown University, who studies how cosmic events shape human history, released a study that showed the 1992 event raised the profile of the World Wide Web, which was just five years old with only 2,700 websites. In an interview with Space.com, Degroot said, 'Fundamentally, the timing was right

A futuristic travel poster for Ceres, touting it as the 'last chance for water until Jupiter' for those heading to the outer solar system. Developed in 2016 for the Jet Propulsion Laboratory by Creative Strategy: Dan Goods and David Delgado.

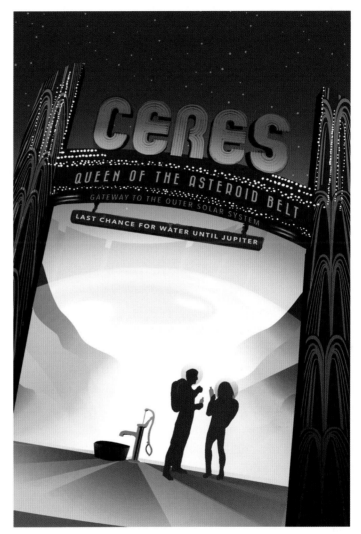

in the sense that the idea that an asteroid or comet could cause a catastrophe on Earth had been well established in popular culture.' Degroot believes that observing the cometary impacts inspired near-Earth asteroid surveys and focused people on the potential benefits of asteroids as resources to be mined. 'It may be that ultimately, this cataclysmic environmental change in a distant corner of the solar system will lead us to alter environments and preserve environments closer to home.'

Names of Asteroids

As John Hind wrote in 1850, 'The most important matter after the discovery of a new planet is the choice of a name.' It was in that year that he was the subject of opprobrium from Benjamin A. Gould, editor of the *Astronomical Journal*, over the naming of the twelfth asteroid. 'Mr Hind', wrote Gould, 'has selected the name Victoria. Such nomenclature is at variance with established usage, and is liable to the objections which very properly led astronomers to reject the name Ceres Ferdinandea.' It is not known how Queen Victoria reacted to having one of her subjects name an asteroid in her honour. W. C. Bond, director of the Harvard College Observatory, came to the defence of Hind when he stated, 'the name appears to fulfill the required conditions of a mythological nomenclature.' Hind offered his rejoinder to Gould. 'It seems to have been forgotten that Her Majesty's name is derived from the goddess, who cannot thereby lose her celestial rights.'

The situation went rapidly downhill from there, hitting rock bottom when Palisa put one of his asteroid discoveries up for auction. He ran an advertisement: 'Being desirous to raise funds for his intended expedition to observe the total solar eclipse of Aug. 29, 1886 will sell the right of naming the minor planet no. 244 for 50 pounds.' That is about £6,000 in modern currency, a tidy sum indeed. In the event the wealthy Baron Albert von Rothschild bought the asteroid name for his wife Bettina. A decade later Edward Holden lamented the nomenclature situation. 'Some of them have a right on a list of heavenly bodies, but many of them, at least, read like the Christian names in a girl's school.'

The convention of naming asteroids for females ended with the discovery of Eros in 1898. It was thought that asteroids crossing the orbit of Mars represented a new type of beast, one well-suited to masculine names. Certain classes of asteroids are given names derived from a specific source. The Trojan asteroids in Jupiter's orbit, for example, are given names from the heroes of Homer's works. It has only been in the last few years that the naming of asteroids has been placed on a firmer footing. For this we have to thank, in a roundabout way, *Star Trek*.

Johann Palisa, who discovered 122 asteroids, put the naming of one of them up for auction in 1886.

In 1985 an astronomer named an asteroid after his cat. He argued that since his cat kept him company during the long hours of observing at night, he had contributed to or been in some way important to astronomy. The cat had been named Mr Spock, in honour of the Vulcan character in the original series of *Star Trek*. While asteroid 2309 was given the name, it began a debate that prompted the IAU to rule out naming asteroids for pets. Names honouring persons, companies or products whose only claim to fame is in the world of business is also discouraged. The actor who played Mr Spock, Leonard Nimoy, was recognized with 4864 Nimoy in 2015 and in 2017, asteroid 31556 Shatner gave due honour to Captain Kirk, William Shatner. No controversy attended these announcements.

Johann Palisa.

The discoverer has the privilege to propose a name, and this suggestion is then accepted or rejected by a Working Group for Small Body Nomenclature of the IAU. It was through a similar process that asteroid 4276, which received its provisional designation as 1981 XA when discovered by Edward Bowell at Lowell Observatory, was named in my honour as Clifford. That was in 1990, when nothing was known about 4276 except its orbit, which is designated as a Mars-crosser (some 13,500 asteroids are now known to cross the orbit of Mars). Subsequent study has revealed 4276 to be 4.4 km (2.7 mi.) in size, with a composition that transitions from an ordinary C-type to the rare B-type asteroids. The most famous B-type asteroid is 2 Pallas; asteroid 101955 Bennu is now the best-studied of all the 65 known B-type asteroids (for the story on how Bennu got its name, see the Appendix).

Music

In 2000 the Berlin Philharmonic commissioned British composer Colin Matthews to write a movement about Pluto to round out the The Planets suite by Holst. Berlin's conductor Sir Simon Rattle also commissioned movements for four asteroids, including Ceres. Kaija Saariaho, Matthias Pintscher, Mark-Anthony Turnage and Brett Dean had the freedom to compose a work on an asteroid of their choice.

Saariaho chose 4279 Toutatis, a potato-shaped Apollo asteroid, for an eerie, shimmering orchestral reflection. Pintscher's Towards Osiris explores the Egyptian myth of reanimation. Delicate effects and gossamer orchestration suggest the beating of Isis's wings. Dean's work is entitled Komarov's Fall, commemorating a Soviet cosmonaut who died upon his return to Earth in 1967; asteroid 1836 is named Komarov in his honour. Turnage tackles a hypothetical impact of Ceres on Earth with glee in a 6 minute 40 second composition. Bold sweeps of colour, a touch of melancholy from the lower strings and warnings from the brass section herald the end of our planet. *Ceres, an 'Orchestral Asteroid'*, premiered in March 2007. As three of the four composers chose to depict cataclysms in the combined 30-minute work of the four movements entitled *Ad Astra*, it is permeated with brio, but the common subject-matter and instrumentation infuse it with conceptual harmony.

In 1953 Gayla Peevy wrote a novelty song, 'I Want a Hippopotamus for Christmas'. In 2018 NASA delivered this gift to everyone on Earth. First seen in 2003, an Aten passed 2.9 million km (1.8 million mi.) from our planet three days before Christmas in 2018. It was the closest approach of this asteroid, 2003 SD_{220}, in three centuries, but it will come even closer in 2070. It has been dubbed the Hippo asteroid, as it looks like the exposed portion of a hippopotamus wading in a river. The object (which has not officially been named) was so close that ground-based radar photos were able to show a ridge, 100 m (330 ft) above the surrounding terrain, that partially wraps around it near one end. Overall the Hippo measures 1,600 m (1 mi.) long. As the largest hippopotamus in the solar system, it has now entered our pop culture. Owing to its orbital characteristics, it may even become a target for a robotic space mission.

FIVE

VISITING AN ASTEROID

Man made his first imaginative trip to the asteroids in 1839. Remarkably, that year saw not one but two visits to Ceres, both published anonymously in London. In *A Fantastical Excursion into the Planets*, a space traveller visits all the planets of the solar system, and takes special delight in his trips to the four asteroids. This tale presaged many of the elements that would feature a century later when people began writing in earnest about asteroids: the shape of the asteroids, whether or not they had atmospheres or inhabitants, and their origin in the explosion of a former planet.

The second visit to Ceres in 1839 is related in a poem entitled 'Immortality'. After arriving at Ceres in only 10 minutes (faster than the speed of light), we are told the surface of Ceres consists of a 'brownish-green turf'. Most amazing of all is that it is inhabited by large numbers of people, thousands of whom immediately crowd around the 'chariot' as it lands. Far from being natives of Ceres, we learn they are cast-offs from Earth!

The first decades of the twenty-first century are a golden age of asteroid exploration. The Voyager spacecraft in the 1970s and '80s gave us our first close-up views of the outer solar system when they flew by Jupiter, Saturn, Uranus and Neptune. The focus has now shifted to exploring the smaller solar system objects: comets and asteroids. Several asteroids, such as 10 Lutetia, have been imaged up close, revealing objects with unusual shapes and battered surfaces.

Montage by Emily Lakdawalla showing all the small asteroids (to scale) visited by spacecraft by 2019 (not included are the large asteroids Vesta and Ceres). Objects not shown at correct relative albedo.

In 2018 alone two spacecraft visited asteroids, and a third ended its mission to explore Ceres and Vesta.

The Dawn Mission to Ceres and Vesta

The most important space mission to the asteroids began on 27 September 2007 when the Dawn spacecraft launched from Cape Canaveral in Florida. It arrived at Vesta on 16 July 2011 and studied the fourth asteroid until 5 September 2012. In a dramatic move, it then left Vesta's orbit and travelled to Ceres, marking the first time a spacecraft ever escaped orbit from one object for the purpose of orbiting another. Dawn orbited Ceres from 5 March 2015 until it ran out of propellant on 31 October 2018. The ability to travel from one world to another was made possible by an ion engine, a technology Dawn's mission director and chief engineer Marc Rayman first encountered in the 1968 *Star Trek* episode 'Spock's Brain'. Rayman

129

Vesta, as photographed by the Dawn spacecraft in this combination of infrared and visible light filters. Vesta is one of the most diversely coloured asteroids that has been imaged. Troughs can be seen running obliquely from the northern shadows to the equator.

was in charge of the Deep Space 1 mission for a comet fly-by in 2001, 'the first spacecraft to use ion propulsion for interplanetary travel'. Pointing to a picture of Vesta in a 2019 video for StarTrek.com, Rayman said 'Vesta is the second largest world between Mars and Jupiter, and my spacecraft was responsible for exploring this alien world. I feel like I'm living *Star Trek*.'[1] Rayman, at the Jet Propulsion Laboratory, California Institute of Technology, describes the importance of the mission:

> The Dawn mission is one of NASA's most remarkable ventures into the solar system. Launched in 2007, the spacecraft completed an extensive exploration of Vesta in 2011–2012 and conducted a spectacular mission at Ceres in 2015–2018. Ceres and Vesta are the two largest and most massive objects in the main asteroid belt. Indeed, of the millions of objects orbiting the Sun between Mars and Jupiter, those two bodies together constitute around 45 percent of the total mass.

A dramatic view of Vesta, taken on 26 August 2011 by Dawn. The detail in this image shows a steep scarp with landslides and vertical craters in the scarp wall. The image has a resolution of about 260 m (850 ft) per pixel.

Vesta has an average diameter of 522 km (324 mi.), small by planetary standards but far larger than most asteroids. Dawn showed it to be more like a mini-planet than like the much smaller bodies most people think of as asteroids. Like a planet, Vesta has a dense core of iron and nickel, surrounded by a mantle, which is in turn surrounded by a crust. The impact, perhaps 1 billion years ago, deep in the southern hemisphere of an asteroid around 50 km (30 mi.) in diameter, excavated a crater more than 500 km (300 mi.) across. At the centre of the crater, a mountain now rises to well over twice the height of Mt Everest above sea level. As Vesta reverberated from the huge impact, the ground broke up far away, near the equator. The resulting system of canyons attests to the powerful punch Vesta sustained.

131

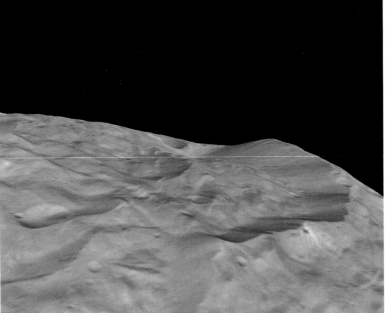

Vesta: this colourful image from Dawn in 2011 shows material northwest of the crater Sextilia. It is located around 30 degrees south latitude, at the bottom right. In this image, the entire colour spectrum of Vesta becomes visible. While a large asteroid impact probably brought the black material, the red material may have been melted by the impact. This composite image maximizes subtle differences in the physical characteristics of rock units, such as colour, texture and composition.

An oblique view of Vesta's south pole reveals cliffs that are several kilometres high, deep grooves and craters. How this wild scenery formed is not yet clear.

Dawn travelled for 2.5 years and 1.5 billion km (900 million mi.) from Vesta to Ceres. At 939 km (584 mi.) in diameter, Ceres is so large that it is included in the category of dwarf planets. (It was discovered 129 years before its better-known fellow dwarf planet Pluto.) Unlike Vesta, which is rocky and mostly dry, Ceres has a large inventory of water, most of it frozen. Dawn provided a wealth

The full disc of Ceres, created with images from Dawn. Occator crater with its bright spots is near the centre.

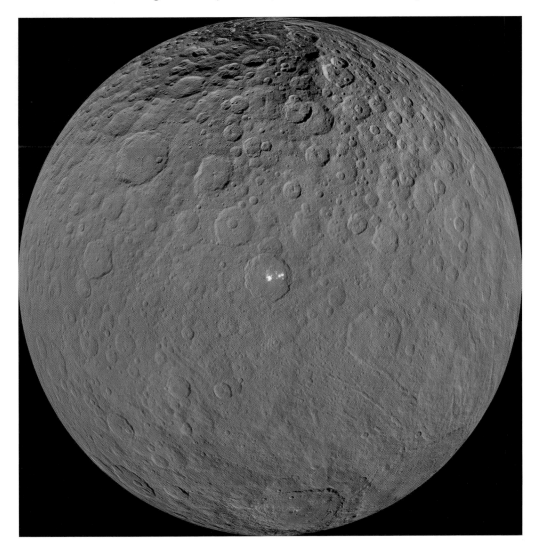

of data that helped scientists conclude Ceres may have been covered in a deep ocean of liquid water early in the solar system's history. That ocean froze and was lost to space long ago, but many of the minerals formed under the pressure of the water and rock back then were detected by the probe's sensors.

Dawn also found almost mesmerizingly bright areas on the mostly dark Cerean ground. When underground salt water made its way to the surface in geologically recent times, the cold vacuum of space caused it to freeze and then sublimate, transforming from a solid to a gas. The water molecules dispersed, but they left the salt behind, making a bright ground covering. Although they shine only by reflecting sunlight, some of these areas are so bright that it is as if they are cosmic beacons. Like interplanetary lighthouses, their brilliant light illuminates the way for a bold ship from Earth sailing on the celestial seas to a mysterious, uncharted port.

Occator crater on Ceres is 92 km (57 mi.) across and 4 km (2.5 mi.) deep. This view, which faces north, was made using images from Dawn, 385 km (240 mi.) above Ceres in 2017.

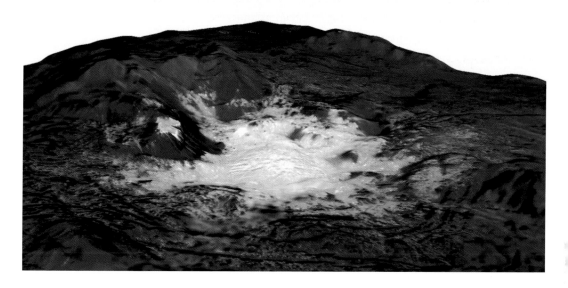

This mosaic of Cerealia Facula at the centre of Occator crater combines images obtained from altitudes as low as 35 km (22 mi.) above Ceres' surface. The mosaic is overlain on a topography model based on images obtained during Dawn's low altitude mapping orbit (385 km altitude). No vertical exaggeration was applied.

Dawn discovered a rich variety of chemicals on Ceres, including organics. It also found evidence that Ceres still has internal heat, left over from the decay of radioactive elements. With all these ingredients, Ceres could experience some of the chemistry related to the development of life. Scientists did not want to contaminate that pristine environment with Dawn's terrestrial materials, so controllers ensured that the spacecraft concluded its expedition in an orbit that would not let it crash for decades or even centuries, allowing time for a follow-up mission if deemed of sufficiently high priority, given the many goals NASA has for exploring the solar system.

Dawn far outlasted the expectations of its human operators. Following an outstandingly successful mission that showed how fascinating, exotic and different two alien worlds in the asteroid belt could be, the adventure ended on 31 October 2018. Although no longer operable, the spaceship will remain in orbit around Ceres. This is a fitting and honourable conclusion to its historic journey of discovery. Dawn's scientific legacy is secure, having revealed myriad fascinating and exciting insights into two complex but quite dissimilar residents of the main asteroid belt. This interplanetary ambassador from Earth is now an inert celestial monument to the power of human ingenuity, creativity and curiosity, a lasting reminder that our passion for bold adventures and our noble aspirations to know the cosmos can take us very, very far beyond the confines of our humble home. (The research was carried out at the Jet Propulsion Laboratory, California Institute of Technology, under a contract with NASA.)

Extensive analysis of the Dawn data has allowed mission scientists to create a topography map of Vesta, and to determine that it was volcanically active for at least 30 million years after its formation;[2] and Ceres's Occator crater, where the bright areas are concentrated, may hold the key to the origin of the largest asteroid. The presence of sodium carbonate and ammonia-bearing salts are indicative of formation near the orbit of Neptune where ammonia

Colour-coded topography map of Vesta constructed from 17,000 images by Dawn. The colour scale extends from 22.47 km (13.96 mi.) below the surface in purple to 19.48 km (12.11 mi.) above the surface in white.

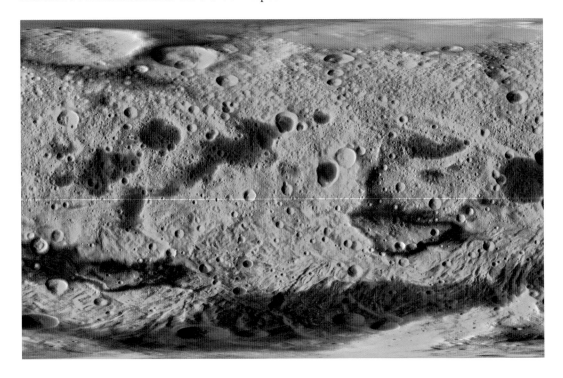

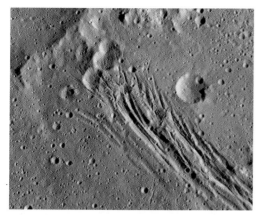

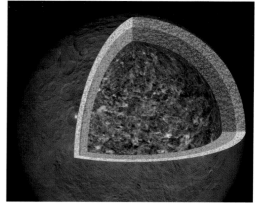

The linear features in this image from Dawn are depressions within Yalode Crater (260 km (160 mi.) diameter) on Ceres. The widest linear feature is 1.5 km across. Photo taken 15 June 2016 at an altitude of 385 km (239 mi.).

This view of Juling Crater on Ceres was constructed from pictures Dawn took from an altitude of 385 km. Juling is 20 km (12 mi.) across.

Artist's conception of the interior of Ceres, created in 2018 based on Dawn spacecraft data.

is abundant. Both substances are in the plumes of material being ejected by Saturn's moon Enceladus, thus linking Ceres with an object that many consider a place where life may exist. Is there life on Ceres, as Oxley wrote in 1877? Certainly not what he envisioned, but microbial life may exist there.

Linear features formed when extensional stresses pulled the surface of Ceres apart. Further evidence of a dynamic Ceres come from successive images of Juling crater: the amount of ice on the shadowed northern wall changed over six months in 2016. Like Earth and other terrestrial planets, Ceres has compositionally distinct layers at different depths, depicted in an artist's conception. The inner mantle (shown as dark green) is dominated by hydrated rocks, like clays. A crust 40 km (25 mi.) thick (shown in white) is a mixture of ice, salts and hydrated minerals. Between the mantle and crust (in brighter green) is a layer that may contain a little bit of liquid rich in salts, called brine. It extends down at least 100 km (62 mi.), which is the deepest Dawn data permits. Hence, it is not possible to tell if Ceres's deep interior contains more liquid or a core of dense material rich in metal.

Sample Return Missions

Of all known asteroids, only five have the optimum orbits, sizes (more than 200 m across) and compositions (carbon rich) for a sample return mission. Two of these are Ryugu and Bennu, both discovered in 1999.

The first sample return mission occurred in 2005 when the Japanese spacecraft Hayabusa took samples of Itokawa. After a drama-filled mission that saw all of its engines fail, the craft returned to Earth in 2010 with more than 1,500 grains of asteroid dust. A study of those grains released in 2019 revealed the presence of water on Itokawa, the first time water was found in samples collected from an asteroid.[3] Similar S-type asteroids, as valuable sources of water, will thus be targeted in future human exploration of the solar system. Hayabusa's successor is now at work: in June 2018 Hayabusa2 arrived at asteroid Ryugu, the name of a mythical underwater castle from a Japanese folktale, in which a young fisherman retrieves a box that, once opened on his return to the surface, transforms him into an old man. Features on Ryugu are now being named, all taken from stories for children. In the control room for the mission sits a stuffed toy: Ryugu-no-tsukai, an oarfish, which in real life can grow up to 11 m (36 ft) in length. Mission controllers love it so much they made a Halloween costume for the oarfish to wear during October 2018.

The journey of Hayabusa2 took 1,302 days, and its initial photos revealed an astonishing world that has raised important questions about asteroid evolution. Measuring 1,000 m (3,280 ft) in diameter at the equator and 875 m (2,870 ft) from pole to pole, it is nearly octahedral in shape: thus it resembles the mineral fluorite, far different from the irregular shape expected before launch. To naturally form that shape it would have had to spin more rapidly in the past, at a rotational rate of about 4 hours, compared to its current period of 7.6 hours. It is also one of the darkest solar system objects ever seen, with an albedo ranging from 1.4 to 1.8 per cent (a typical C-type asteroid is 3 to 4 per cent). Further research will try to figure out why it is darker than any carbonaceous chondrite meteorite we have studied on Earth. Composition of the asteroid indicates it once interacted with water, a characteristic it shares with many meteorites. Scientific data released in April 2019 indicate Ryugu is a rubble pile asteroid, as evidenced on its surface, which is literally covered in boulders. One is so big, 160 m (525 ft) across, that it has been given its own name, Otohime; it sits near the south pole.

The largest crater on Ryugu is 220 m (720 ft) across, but most craters measure just 1 to 30 m in size. There are no smaller craters on the surface because impactors smaller than 0.1 m will just fracture existing boulders. A study of the craters on Ryugu indicate they are no more than a million years old, as more ancient ones get erased by subsequent impacts and a process termed 'seismic shaking'. This happens when the asteroid gets hit by an object large enough to trigger asteroid-wide shaking, dislodging boulders that settle, eventually filling in the craters.

The boulder dubbed Otohime on the asteroid Ryugu is 160 m (525 ft) wide. It sits inside a deep furrow that winds nearly completely around the asteroid.

2018-07-20 07:12:30 UTC

Ryugu from a distance of 6 km (3.7 mi.) as imaged by the Japanese spacecraft Hayabusa2 on 20 July 2018. This shows its largest crater, Urashima.

On 22 February 2019 Hayabusa2 landed on Ryugu for just one minute. The team at the Japanese space agency made a statement harkening back to 1969 when man landed on the Moon: 'One small hand of mankind has reached for a new, little "star".' The craft shot a tantalum bullet into the surface at a velocity of 300 m per second. This was designed to kick up dust, which then fell into a collecting tray that will be returned to Earth for analysis.

Hayabusa2 made history on 4 April 2019 by bombing an asteroid for the first time. The 2 kg (4½ lb) copper explosive charge sent to the surface at 2 km per hour (1.2 mph) made a crater; in July the craft landed in the crater to collect underground samples. Hayabusa2 also made history by sending the first rovers to an asteroid. Two rovers (dubbed Owl and Hibou) landed in September 2018, with a

Artist's impression of Hayabusa2 as it made its touchdown on Ryugu in 2019.

third sent to the surface on 3 October 2018. The size of biscuit tins, they checked out both the February 2019 landing site and the April 2019 explosive site in advance. A fourth and final rover was sent to the surface in October 2019. The rovers jumped along the surface of Ryugu as the very slight gravity of the asteroid precluded surface travel. Hayabusa2 departed the asteroid on 13 November 2019 and is scheduled to return to Earth with samples in 2020.

At the same time as Hayabusa2 studied Ryugu, an American space mission dubbed OSIRIS-REX orbited asteroid Bennu, a 500-metre-wide (1,640 ft) body. OSIRIS-REX, launched in 2016, went into Bennu's orbit on 3 December 2018. This is not any ordinary orbit: it is in fact the closest any spacecraft has orbited a celestial body, just 1.19 km from the surface. Based on thirteen years of

The OSIRIS-REX spacecraft launches aboard a ULA Atlas V 411 rocket from Cape Canaveral Air Force Station, Florida, on 8 September 2016.

This mosaic image of asteroid Bennu is composed of twelve images collected on 2 December 2018 by the OSIRIS-REX spacecraft from a range of 24 km (15 mi.). The image was obtained at a 50° phase angle between the spacecraft, asteroid and Sun, and in it, Bennu spans approximately 1,500 pixels in the camera's field of view. Bennu's tallest boulder, named Benben Saxum, juts out 44 m (140 ft) from the southern hemisphere, at bottom right.

observations, astronomers revealed in 2019 that Bennu is speeding up its rotation period by one second per century. This YORP effect has been detected in other asteroids, but the probe is in a unique position to measure it independently of telescopic data. Dante Lauretta, prime investigator on the Bennu misson, said 'In about a million and a half years we predict that Bennu will be spinning at twice its current rate.' Its shape, similar to Ryugu, may also be the result of its spin rate.[4] The YORP effect may not result in a linear progression of increased spin rate, however, as there is some question as to why Bennu is not spinning faster, so other factors may be at work.

The prime mission of the spacecraft is to gather surface samples in August 2020 at the Nightingale landing site, and return them to Earth on 24 September 2023, but in the meantime it has already made headlines. In a seminar at the University of Arizona in late 2019, science team chief Michael C. Nolan said, 'Bennu doesn't match any known meteorite spectra, and may represent a new type of meteorite never seen before. In our investigation of Bennu we are finding something new and unexpected that's going to teach us about how the Earth formed, and how the solar system formed.' Another surprise is the presence of bright rocks on the surface, with spectra that match that of Vesta; the overall surface of Bennu has an extremely dark albedo of 4.4 per cent.

In a press conference when scientific data was released in 19 March 2019, Lauretta said that one of the biggest surprises of her scientific career happened when OSIRIS-REX detected particles that are ejecting off the surface at relatively high velocities, so Bennu is one of a very small group (a dozen to date) of small bodies called 'active asteroids'. Several events witnessed have ejected hundreds

of particles in the size range of a centimetre to tens of centimetres. Some are being ejected at several metres per second, which means they are being ejected into interplanetary space. Some of the slow-moving particles appear to be trapped in the asteroid's gravity field, meaning it is creating its own set of natural satellites; others have been observed to fall back on its surface. Basically, stated Lauretta,

The images creating this mosaic of the site where OSIRIS-REX landed on 20 October 2020 were collected on 26 October 2019 from a distance of 1 km (0.6 mi.). For scale, the large boulder in the centre of the image is 14 m (46 ft) wide.

This view of asteroid Bennu ejecting particles from its surface on 19 January 2019 was created by combining two images taken by OSIRIS-REX: a short exposure image (1.4 ms), which shows the asteroid clearly, and a long exposure image (5 sec), which shows the particles clearly.

'it looks like Bennu has a continuous population of particles raining down on it from discrete ejection events across its surface. This is incredibly exciting – we don't know the mechanism causing this.'

The largest particle ejection observed, on 6 January 2019, 'looked like an open star cluster off the edge of the asteroid', said Carl Hergenrother on the OSIRIS-REX team, who co-authored a study on plausible mechanisms to explain this activity.[5] The sample return concept for OSIRIS-REX differs from that of Hayabusa2. The American craft has a nitrogen jet at the end of a robotic arm. This gas will disturb the surface sufficiently to blow particles into the sample collector. The head of the collector is also covered in Velcro-like pads designed to pick up surface dust on contact. The maximum Hayabusa2 will deliver to Earth is 100 mg, the weight of three grains of rice; the larger American craft is designed to capture

2 kg, 20,000 times as much. In the spirit of cooperation, the space agencies will be sharing the spoils: 10 per cent of the Ryugu sample will go to NASA, while 0.5 per cent of the Bennu sample will go to JAXA.

Future Spacecraft Missions

Spacecraft visits to asteroids may be broadly separated into two spheres of activity. In the first, humanity sent probes to fly-by, orbit or land on asteroids to gain scientific knowledge. In the second, the concept of planetary defence will be put into practice, with pure science results taking a secondary role. I write 'will be' because it has not happened yet, but will happen quite soon. The idea of mitigating asteroid impacts on Earth, which are potentially devastating, has been around for decades. I attended a conference on the subject, held in San Juan Capistrano in California. That was way back in 1991. A spacecraft slated for launch in 2021 will finally address the problem head-on.

The craft is dubbed DART (short for Double Asteroid Redirection Test). Embedded in this title are two key concepts. First, double asteroid. The target of DART is the NEO 65803 Didymos (Greek for twin), which is a binary. This will mark the first visit to such an object. But more importantly for this discussion, it will also be the first kinetic impact to deliberately target an asteroid with the purpose of redirecting it. That is not to say that Didymos, 780 m (2,560 ft) in size, is a threat to Earth. In fact it is not considered an imminent threat at all; rather, it is being used a test subject. More precisely, its smaller component, the moon Dimorphos, will be hit by DART.

Just hitting the target will require great precision, as the craft's camera will not see the disc of the moon until ten hours before impact. Photographs shot as it hurtles towards Dimorphos may let us see objects on the surface as small as 1 cm across. When the probe crashes into the asteroid at 6 km per second, its demise

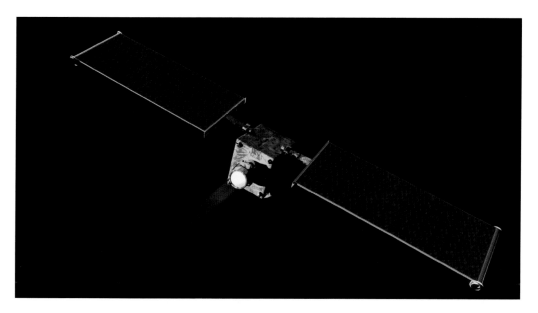

Artist concept of NASA's Double Asteroid Redirection Test (DART) spacecraft. DART, due for launch in 2021, will be NASA's first mission to demonstrate an asteroid-deflection technique for planetary defence.

may be imaged in real time by a miniaturized satellite known as a cubesat, cutely dubbed SelfieSat. The timetable at time of writing is for impact in October 2022, which could be a banner year for asteroid studies. Another cubesat, NEO Scout, is slated for launch by NASA in 2021. It will use a solar sail to visit one or more NEOs.

The European Space Agency (ESA) is also planning a mission to Didymos, which will give us a wealth of data on the nature of a binary asteroid. Of special interest will be a view of the crater caused by DART and how that impact altered the orbit of the tiny asteroid moon Dimorphos, only 160 m (525 ft) across. This data will be used to refine our planetary defence strategy. The launch of the ESA's Hera is set for 2024, with arrival at the asteroid in 2027. If all goes well it will be the pathfinder for a new kind of spacecraft that will be crucial for twenty-first-century study of asteroids. This is because, similar to a self-driving car, Hera is capable of autonomous navigation. Once its main objectives are completed, with the usual ground-based operations from Earth, Hera will switch over to sensing its own environment and navigating as it sees fit.

Near-Earth Asteroid Scout (NEA Scout) is a 2021 mission by NASA to analyse, design, develop and fly a controllable cubesat solar sail spacecraft capable of encountering near-Earth asteroids (NEA).

Hera spacecraft at Didymos in 2027, shown in this artist's concept with two cubesats it may deploy to test intersatellite communications in deep space.

A NASA spacecraft dubbed Lucy, slated for launch in 2021, will visit several Jupiter Trojan asteroids, including Patroclus in 2033, while a Japanese solar sail mission dubbed OKEANOS envisions a sample return to Earth of a Trojan asteroid in the 2050s. Another NASA probe, launching in 2022, will orbit the largest metallic asteroid of the solar system in 2026: that is 16 Psyche, discovered in 1852, a chunk of iron, nickel and various rare Earth metals. Psyche is almost certainly the exposed core of a differentiated object. Whatever material originally surrounded it has been blasted away by impacts with other planetesimals or asteroids, revealing a core that was still molten when it was exposed to space. Research released in April 2019 raises the high likelihood Psyche has been subject to iron volcanism: the melt contained within the crust which formed after exposure is heavier than the molten iron beneath, which can erupt through any conduits that reach the surface. A study of metallic meteorites might yield evidence for such iron volcanoes. The Psyche space mission will thus give us the first look at an iron core similar to those that exist inside terrestrial planets, including our own.[6]

Plans are also afoot to send a small satellite to Pallas, piggybacking on the rocket that is going to Psyche. Since expensive spacecraft cannot be sent to every asteroid of interest, a strategy is developing to send many small satellites (smallsats) to do the job. While these would just be fly-by missions, they would be able to give us much basic science data and photographs of potentially hundreds of asteroids. The cost of a smallsat is one-tenth that of a flagship mission like Dawn, which cost $467 million. The probe to Pallas is called Athena: the ancient Greek goddess was often referred to as Pallas Athena. A mission to Pallas is certainly long overdue as this significant world remains unexplored, while several tiny asteroids have been thoroughly studied. The fly-by is planned for 2023.

Engineers at the University of Arizona proposed an asteroid mobile imager in 2019 that will take the concept of placing smallsats

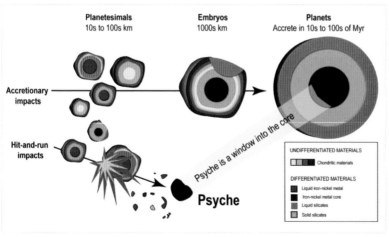

Planetesimals
10s to 100s km

Embryos
1000s km

Planets
Accrete in 10s to 100s of Myr

Accretionary impacts

Hit-and-run impacts

Psyche is a window into the core

Psyche

UNDIFFERENTIATED MATERIALS

Chondritic materials

DIFFERENTIATED MATERIALS

Liquid iron-nickel metal
Iron-nickel metal core
Liquid silicates
Solid silicates

Artist's conception of a NASA mission to asteroid Psyche, planned for 2026.

The metallic asteroid Psyche offers a unique window into the violent history of collisions and accretion that created the terrestrial planets.

on the surface of an asteroid. These autonomous semi-inflatable robots designed to operate in a swarm will represent a novel approach to studying the surface of a large number of asteroids. They will essentially be scouts, giving us an understanding of a target asteroid in preparation for larger and more sophisticated landers. Once deployed from a mother ship, each tiny lander (known as AMIGO) is designed to inflate to a 1 m sphere, with a camera on top. With the use of intelligent control algorithms, it will be able to hop along the asteroid's surface with the use of cold gas micro-thrusters and a reaction wheel.[7]

The first delivery in 2019 of a solar electric propulsion thruster called NEXT will be a key factor in enabling asteroid missions in the future. Three times as powerful as the thruster used by the Dawn spacecraft, development of NEXT at NASA's Glenn Research Center took a decade. Thousands of asteroids will be within reach of NEXT thrusters powering motherships carrying smallsats.

China announced plans in 2019 for an asteroid sample return mission that might launch in 2024 to asteroid 469219, which has the Hawaiian name Kamo'oalewa. This object, found in 2016, has an unusual orbit and is classed as a quasi-satellite of Earth. It is cutely known as Earth's 'pet rock'. Its maximum distance from Earth is one hundred times the distance to the Moon. The mission also calls for the spacecraft to visit comet 133P/ Elst-Pizarro, which is also classified as an asteroid. Perhaps one day a human will visit an asteroid, but visiting any except the largest will be a major technological challenge as their gravity is so low. The concept of establishing human colonies on asteroids will surely remain in the realm of science fiction.

A 2017 painting by Jim Scotti of the Lunar and Planetary Laboratory of a future astronaut visit to an asteroid.

Mining

In 2015 the President of the United States signed a space-mining
bill passed by Congress. 'A United States citizen', it reads in part,
'engaged in commercial recovery of an asteroid resource under
this chapter shall be entitled to any asteroid resource obtained,
including to possess, own, transport, use, and sell the asteroid
resource.' Further, the bill encourages the commercial exploration
and utilization of resources from asteroids.

    The idea of capturing an asteroid sounds like science fiction,
but it was official United States policy for several years. In 2010
President Obama approved NASA's plan to send a robotic spacecraft
to an asteroid, lasso it and send it into lunar orbit. When it finally
reached the NASA budget request, submitted to Congress in April
2013, NASA administrator Charles Bolden said, 'We are developing
a first-ever mission to identify, capture and relocate an asteroid.
This asteroid initiative brings together the best of NASA's efforts
to achieve the president's goal of sending humans to an asteroid in
2025.' The total cost of the Asteroid Redirect Mission was estimated

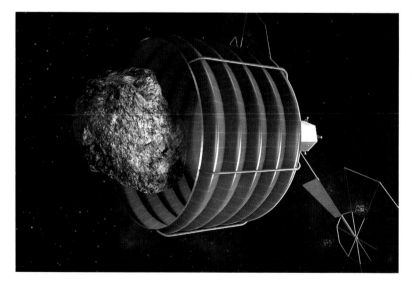

Notional concept of a
solar-electric-powered
spacecraft, designed to
capture a small NEA and
relocate it safely close to
the Earth-Moon system so
astronauts can explore it.

at $2.6 billion. The budget sent to Congress four years later, in April 2017, removed funding for the project, but the solar electric propulsion system being developed for the mission was completed in 2019.

The prototype for asteroid mining began in 2019 when the Hayabusa2 spacecraft fired a high-speed projectile into the surface of 162173 Ryugu. This 'shoot and collect' method is one option for space mining. Because of their low gravity, landing people or even mining equipment on most asteroids is not possible. But a craft could get very close to a small asteroid and get a lot of material by shooting it with projectiles, causing material to leave the surface where it could be collected. Another option being explored in 2020 by the NASA Innovative Advanced Concepts programme is optical mining, whereby sunlight concentrated onto the surface would generate debris to be collected into an inflatable bag.

Future astronaut scientists survey the surface of an asteroid that has been 'pulled' into Earth's orbit to study its compositional type and possible mining opportunities. Original art for Bantam Books, 1991. Painting by Pamela Lee, whose artwork has flown on the American Space Shuttle and the Russian Mir space station.

The news media in February 2013 was filled with stories about asteroid 2012 DA_{14}, a 45-metre (148 ft) rock that made a particularly close approach to Earth. Bradford Space (formerly Deep Space Industries), an American firm that plans to pioneer asteroid mining, calculated its value on a presumed weight of 130,000 tonnes. The figures were startling: water worth U.S.$65 billion, and (assuming it consists of just 10 per cent metals) another $130 billion in metals such as iron and nickel.[8] It has been suggested that the first trillionaire will be an asteroid miner. Even asteroids are not infinite in resources, though, leading a scientist and a philosopher in 2019 to call for humanity to adopt the 'one-eighth principle', which

A conceptual look at an asteroid mining mission to a near-Earth asteroid, based on a NASA-sponsored study on space manufacturing held at Ames Research Center in 1977. 'Asteroid-1' is the central long structure and the propulsion unit is the long tubular structure. Solar cells running the length of the propulsion system provide power. In the left foreground is an asteroid mining unit. An orbital construction platform in orbit provides infrastructure for the project.

would see only an eighth of solar system resources made available to exploitation, the remainder to be preserved as wilderness. 'One-eighth of the iron in the asteroid belt is more than a million times greater than all of the Earth's currently estimated iron ore reserves,' write Martin Elvis and Tony Milligan, 'and it may well suffice for centuries.'[9]

Tsiolkovsky realized that asteroids were sources of metals that could be mined, but he likely never even considered the possibility that a lifeless rock might contain water, a commodity of the utmost importance for deep space missions. We now know, writes planetary scientist Don Yeomans, that their 'hydrated, or clay, minerals and ices can be broken down into hydrogen and oxygen, the most efficient form of rocket fuel'. In 2019 NASA initiated funding of a project at the Colorado School of Mines to mine ices on cold solar system bodies for exactly that purpose. David Gump, the chief executive of Deep Space Industries, stated in a press conference in 2013 that 'even with conservative estimates of the potential value of any given asteroid, if we begin to utilize them in space they are all the equivalent of a space oasis for refuelling and resupply.'

APPENDIX:
ENGAGING WITH ASTEROIDS

Asteroids have been put to many uses, from probes of the origin of the solar system to opportunities for space mining, but using them as a test-bed for studying gravity must rank as the most unexpected. In 1823 the German astronomer Friedrich Bernhard Nicolai postulated that the attraction of Jupiter on the asteroid Juno was different from that upon the Sun. Thus he came to the conclusion that gravity acted differently on various celestial objects. It was not until the 1830s that careful measurements by George Airy uncovered an error in Nicolai's work. A book of 1909 summed it up nicely: 'The ideal beauty of these planets [the asteroids] is that their reappearance is continually a test of the correctness of mathematical assumptions and that they exhibit over and over again the validity of Newton's law of the universal attraction of all bodies.'[1]

In 1871 the Swedish astronomer Hugo Gyldén made an important advance in our understanding of the galaxy. He likened the presumed rotation of the galaxy, which had not been proven yet, to be like the planets revolving around the Sun under the action of gravity: those stars closest to the hub would revolve quickly, while those in the distant reaches would move slowly. He tested this theory by using the motions of asteroids. We know they revolve around the Sun, but just suppose one were presented with the data on asteroidal motion in the absence of this knowledge. Could one then establish that they were in fact orbiting a central hub of some description? He found the answer was affirmative: they do indeed orbit a central point in a direction just 6 degrees away from the Sun. By applying that same reasoning to the stars of the galaxy, he identified a line running through its centre. But which way would the centre lie? Was it in Sagittarius or 180 degrees away in Taurus? Gyldén unfortunately chose Taurus, but this was a truly extraordinary use of asteroids to extend our knowledge of the universe.[2]

The minuscule gravity of asteroids has inspired thoughts of sporting events. Edmund Ledger, in 1895, told an audience at Gresham College in London that he 'could not say if we should jump so high as to go out of sight, but with a good jump we should be absent for about an hour before we came down again, which fact presented the idea of very interesting athletic contests on the minor planets.'

Today the search is on for planets orbiting other stars, and since each planetary system formed under the same laws of physics (including gravity) as ours, it seems certain that these other systems resemble ours in general. 'The firmament is a majestic, awesome sweep of stars, somewhat as you see it from Earth! And there are other worlds, moons, meteors, comets and asteroids, just as you'll find in our own solar system.'[3] This quotation comes not from a description of what a space telescope of the 2030s will almost certainly see, but from a comic book written in 1961.

A particularly fascinating application of gravity on an asteroid will take place in April 2029 as Apophis makes a very close approach to Earth. Our planet's gravity will not only change the orbit of Apophis but may change its spin rate and trigger avalanches on its surface.

Past Projects

Before having a look at what you can do in the future to engage in asteroid studies, let's look back to the early 1990s, in the pre-CCD era. Jordan Marche, professor of astronomy at the University of Wisconsin-Madison, remarks about his work with a telescope he made himself.

> As for the asteroid work, I had written a plate reduction program in BASIC that I shared with Roger Sinnott at *Sky and Telescope*. This was somewhat in response to a column that Brian Marsden had written in 1988 on 'What Amateurs Should Be Doing'. Sinnott published our 'hybridized' plate reduction program in S & T, July 1990, 71–75, but attributed it entirely to me, which is incorrect. I afterwards corrected a faulty assumption in my program and used it subsequently with my own photographs taken on film.

A decade earlier, in the early 1980s, I achieved accuracy of 0.01 magnitude on asteroid lightcurves with my 36.2 cm reflector equipped with an analogue photoelectric photometer. Thus it was possible in the 1980s and '90s to obtain not only physical data such as rotation periods with photometers, but orbital data from images of asteroids, without access to professional observatories.

The value of back-garden observatories to asteroid work is immense. Near Tucson, Arizona, Roy Tucker made many discoveries. Tucker worked as an engineer in the Imaging Technology Laboratory of the University of Arizona, and an instrument technician at Kitt Peak Observatory. The 1950s and '60s were an exciting time for astronomy and inspired Roy's interest in astronomy and solar system exploration as a young boy – he received a 3-inch refractor at Christmas in 1966. His first astronomical endeavour involving asteroids was a successful collaboration with

Jet Propulsion Laboratory astronomer Alan Harris to observe the occultation of a star by 3 Juno on 11 December 1979. Throughout the 1980s and early '90s Tucker was employed in various technical instrumentalist positions (including instrument support specialist at the Multiple Mirror Telescope), but did not secure a permanent residence and steady job until 1995. The following year he built his own observatory and readied the instrumentation to pursue his passion of asteroid hunting. Tucker described his astronomical journey to me in the following way:

> My efforts to find unknown asteroids began early in 1997 with a Celestron C14 telescope and a CCD camera of my own design and construction that used a 512 × 512 pixel SITe TK512 imager. This combination yielded a field of view of only 12.15 arcminutes but permitted me to image stars and asteroids as faint as magnitude 20.5. The procedure was to acquire triplets of images separated in time by about twenty minutes and then blink them using a program called Astrometrica. The process was very slow and tedious and permitted searching only a very few square degrees per night. Even so, by the middle of June I had discovered seven main belt asteroids. Then on June 28th, I found an asteroid that made a streak on my images, obviously moving quite fast and likely near the Earth. Further observations revealed this to be an Aten-class near-Earth asteroid (NEA), only the 25th known at that time and the first discovered by an amateur.

Encouraged by his early success, Tucker, during the summer months, made modifications to his camera to use a 1024 x 1024 SITe TK1024 CCD, which almost quadrupled his field of view. He added,

> During the fall and winter of 1997, I added another eleven asteroid discoveries. Almost at the end of the year, in December,

I implemented scan-mode imaging which greatly increased my sky coverage and efficiency on operation. (In scan-mode imaging, there is no shutter and the process is continuous from evening twilight to morning twilight as a long strip of sky is imaged. The light is being collected at the same time the electronics are reading out the accumulated charge and the field of view is drifting along to the east.) I also began using an image examination and astrometry program called 'Pinpoint' by Robert Denny. It included a moving object detection capability that could assist me. At this time, I was essentially emulating the SkyWatch Camera but with an aperture of only 14 inches instead of 36 inches. During 1998, I was credited with the discovery of 29 asteroids including an Apollo, 1998 FG2, and another Aten, 1998 HE3. I was also the co-discoverer of comet P/1998 QP54. My scanning would also yield known asteroids and asteroids that had been previously observed but not yet attributed to a discoverer. The Minor Planet Center assigns discovery credit by determining which set of observations most closely describe the actual orbit. On any good night, I might see half a dozen new objects and many known asteroids. Perhaps the most productive night while operating in this manner was October 5th, 1999. On that night, I found nineteen unknown objects of which I was credited with seventeen as the discoverer.

He continued in this way until the end of 1999, when he decided it was necessary to take the next step and implement an array of three telescopes

that would permit the acquisition of image triplet in a continuous, uninterrupted stream from evening to morning twilight. Construction of the system occupied all of 2000 and the early months of 2001. The instrument, called MOTESS for 'Moving Object and Transient Event Search System', began operating in May of 2001 and permitted, depending upon time of year, imaging of 100 to 140 square degrees per night to magnitude 20.5. During the period from December 1996 to April of 2001, I was credited with 99 asteroid discoveries. During the next nine years, I discovered 610 asteroids, including three NEAs and another comet. The operation of MOTESS also resulted in the discovery of tens of thousands of variable star candidates and the pre-discovery photometry of a supernova. The instrument has been very productive.

MOTESS ceased looking for asteroids in early 2010. The professional surveys had just become too good and I was finding very few unknown objects. The exciting 'Golden Age' of amateur asteroid discovery was drawing to a close. MOTESS is now dedicated to variable star observing and has a few more years of life before obsolescence and retirement catches up with it. It has been a joy to operate it.

Tucker's dedication produced 709 numbered asteroids including an additional three NEAs, making him the 24th most prolific asteroid discoverer, amateur or professional, as of 2020.

Future Projects

What is citizen science? As explained in 2019 by Gary Morris of St Edward's University in Austin, it 'involves harnessing the power of small efforts by thousands of people to contribute to studies of some big local and global problems: climate change, biodiversity, land use and asteroid hunting'.[4] New technology is now giving a second lease of life to citizen astronomers who want to add to our knowledge of asteroids, and with the August 2019 discovery of an interstellar comet with a home-made 0.65 m telescope, the opportunities for such dedicated astronomers is vast. As mentioned in Chapter Two regarding the white dwarf star, the detailed study of the star (dubbed LSPM J0207+3331) was made by the Keck II telescope in Hawaii, but that giant telescope only looked at the star because of a citizen scientist. NASA's Wide-field Infrared Survey Explorer (WISE) mapped the entire sky in infrared, generating far more data than professional astronomers have time to study. The data from the WISE mission has been placed online for citizens around the world to study. One such person is Melina Thevenot of Germany, who was searching for brown dwarf stars, but since this particular object is too bright and distant to be a brown dwarf, she brought it to the notice of the WISE project team in October 2018. This has led to the realization that J0207 may be surrounded by several rings of asteroids. Yes, asteroids 145 light years away! Those interested in participating can join the Backyard Worlds: Planet 9 project, led by Marc Kuchner at NASA.

Those with backyard instruments of modest size (a 100 mm refractor or a 6 in. reflector) can typically see a dozen asteroids annually that reach magnitudes brighter than 10. A star chart

On the right in this artist's conception is a debris disc consisting of asteroids and small particles surrounding a dwarf star. An exoplanet, to the left, shares its space with more asteroids.

will be useful to plot the changing location of the asteroids from one observation to another. A much more challenging task is to observe asteroids passing in front of stars, an event known as an occultation. Faint asteroids typically occult faint stars. However, the brightest star in the sky, Sirius, was occulted by an asteroid on 18 February 2019. This was the first ever predicted occultation of Sirius, but it remained unobserved owing to clouds and technical reasons. Occultations, which last only a fraction of a second, can reveal information about the size of the asteroid.[5] Those interested in chasing these rare events should begin by consulting the website www.occultationpages.com.

A new way to observe occultations and develop physical information about asteroids became available in 2019 when the Unistellar telescope became available for sale in Europe and North America. The asteroid scientist Franck Marchis explained its utility for asteroid studies to me during a meeting at the annual megaconference South by Southwest in Austin, in March 2019. Users can download a phone app for instant notification of newly discovered asteroids found by asteroid survey programmes such as Pan-STARRS and the Catalina Survey. Once the position of the asteroid is transmitted to the telescope, it will gather photons in that area of the sky. With a field of view the size of the Moon, it should be able to identify a moving object down to magnitude 16, or even magnitude 18 in a dark sky setting. Such a rapid response by potentially dozens or hundreds of telescopes worldwide will greatly enhance our ability to track them long enough for an orbit determination. Unistellar will also be able to observe asteroid occultations with 3 millisecond accuracy, which has the potential of giving us data on the period and shape of asteroids, as well as finding new moons orbiting them. Astronomers operating a Unistellar were among a dozen teams who observed an occultation of a star by the Jupiter Trojan 11351 Leucus on 29 December 2019. This asteroid is being targeted for a fly-by in 2028 by the Lucy spacecraft.

Aside from the purely scientific aspects, looking at an asteroid pass close to Earth is a lot of fun (as long as it doesn't hit us!). An unusually fine opportunity came on 22 March 2019, when 2019 EA2 passed closer to Earth than the Moon (a 'miss distance' of just 0.002 AU). Since it moved only 5 km per second, slower than most NEOs, it was easier to spot, and with a size of 39 m (128 ft) it was larger and brighter (mag. 15.9) than most NEOs. In comparison, 2019 EA2 is twice the size of the asteroid that exploded as a bolide meteor over Russia in 2013, but about the same size as an asteroid that hit the Moon during a total lunar eclipse on 20 January 2019. For those who don't have a Unistellar and its bot that searches for newly discovered NEOs, a telescope of at least 20 in. will be required to see most such asteroids, but one opportunity for smaller telescopes is viewing or taking photos of asteroid conjunctions. For example, on 5 July 2014 Ceres and Vesta came within 10 arcminutes of one another, just a third of the angular size of the full Moon disc.

The mission to Bennu has set a new standard for how the public can interact with an asteroid space mission. The fascination with asteroids is widespread, and one that the researchers behind OSIRIS-REX were eager to embrace. The University of Arizona, in collaboration with the Planetary Society, held a 'Name That Asteroid!' contest in 2013 that resulted in acceptance of the name Bennu. The winning entry was from a nine-year-old in North Carolina, Michael Puzio. Since the spacecraft is named after an Egyptian god, Osiris, it was deemed fitting that its target asteroid be named for another Egyptian deity. Several such deities were offered among the 39 semi-finalists, but the name associated with the heron deity won out. Bruce Betts, Director of Projects for the Planetary Society, said:

The name 'Bennu' struck a chord with many of us right away. While there were many great entries, the similarity between the image of the heron and the TAGSAM arm of OSIRIS-REX was a

Students as part of the Sudbury Field School exploring a roadside outcrop of granophyre, the upper parts of the Sudbury Igneous Complex, the ~3-km thick melt sheet generated by the impact of an asteroid into what is now Ontario, Canada.

clever choice. The parallel with asteroids as both bringers of life and as destructive forces in the solar system also created a great opportunity to teach about planetary science.

An innovative way to name asteroids was used in 2019 for the largest unnamed object in the solar system: a public vote. The three astronomers who discovered 2007 OR_{10} came up with three suggestions for a name, associated with mythological creatures and figures that reflect aspects of its physical properties, which they posted online. With 280,000 votes cast, the winner was Gonggong, a Chinese water god. The name was formally adopted in 2020.

On a more substantive note, NASA mobilized citizen scientists in 2019 to help select a landing spot for OSIRIS-REX by creating a hazard map. This consists of measuring Bennu's boulders and mapping its rocks and craters via the web interface CosmoQuest, a project run by the Planetary Science Institute. Similar opportunities to get actively involved in a space mission will certainly happen in the future.

Those who are really serious about NEOs can apply for a Shoemaker NEO Grant from the Planetary Society, which offers funds for astronomers interested in physical studies and astrometric follow-up. Grants are typically $5,000 to $12,000. The award programme is named in honour of the great planetary scientist Eugene Shoemaker.

The adventurous who want to get their hands dirty can turn to the Centre for Planetary Science and Exploration. They have just the event for you: an Impact Cratering Short Course and Field School. It is led by Gordon Osinski at the University of Western Ontario in Canada, who explains:

Meteor Crater in Arizona. The hummocky deposits just beyond the rim are remnants of the ejecta blanket. This aerial view shows the dramatic expression of the crater in the arid landscape.

> This is an intensive 7-day short course and field training program on the processes, products and effects of impact

cratering. I start by introducing students to the theory and the concepts in the classroom and then we move into looking at the products, which are more tangible. Between going out into the field in Sudbury and the rock kits that I have put together, the students get to learn in a hands-on environment, which as a teacher, I believe is the best environment. This course is based in Sudbury, Ontario, the site of an ~200 km diameter impact structure formed 1.85 billion years ago and it offers an exceptional opportunity to study impact melt rocks, various types of impact breccias, shatter cones, impact-induced hydrothermal alteration, and much more.

Osinski began teaching the course in 2009. 'I've had 116 students take it so far, from Canada, the USA, Europe and even one student from Japan.'

Remember, if you hold a nickel coin that is actually made of nickel, there is a good chance it came from an asteroid: the Sudbury

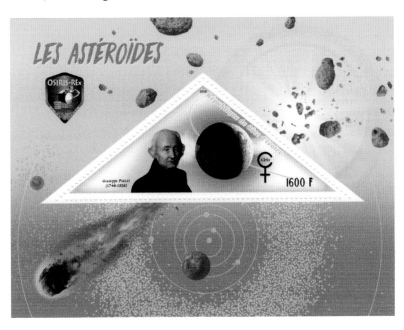

A 2018 postage stamp from Ivory Coast depicts Giuseppe Piazzi and the first asteroid, Ceres. At the upper left is the logo for the OSIRIS-REX asteroid mission to Ceres.

Igneous Complex is a rich source of nickel, copper and platinum group metals derived from an asteroid. While you can't see an impact crater there, a spectacular one can be visited in Arizona where Meteor Crater is a major tourist attraction. Guided tours around the rim of the crater are available daily; I had the good fortune to be led on a field trip to the bottom of the crater by Shoemaker, who was the key figure in understanding the crater was caused by a meteor impact 50,000 years ago. It is 59 km (37 mi.) east of Flagstaff, where Percival Lowell spent several years searching for the ninth planet. The Pluto Discovery Telescope, which I used while working at Lowell Observatory, is now on public display after a restoration effort in 2018. It is worth a visit to see the first telescope that ever photographed a TNO. That was back in 1930 when Clyde Tombaugh discovered Pluto at Lowell, at that time under the direction of Vesto Slipher, who deserves credit for continuing the search and hiring Tombaugh.

One inexpensive and unusual way to get engaged in asteroid-related activities is to collect postage stamps, coins and medals that feature asteroids from countries all over the world. For example, the Japanese mint issued a medal in 2010 to commemorate the first Hayabusa spacecraft. To get up close and personal with space rocks, visit museums that have meteorite collections; the largest collection of lunar and Martian meteorites opened in Bethel, Maine, in December 2019. Searching for meteorites, or merely collecting them, is another way to engage with asteroids. Artists are also crucial in enhancing interest in the subject for the public. Have a look again at the painting of Patroclus and its moon shown in Chapter Two. Artist Lynette Cook shared her creative process:

> The illustration was a commission for the press release from the Keck Observatory. I worked directly with Dr Marchis to find out what he felt the art should show. My first step was to create some roughs for review, then I took the best one and made revisions until he gave me the 'thumbs up.' At that point

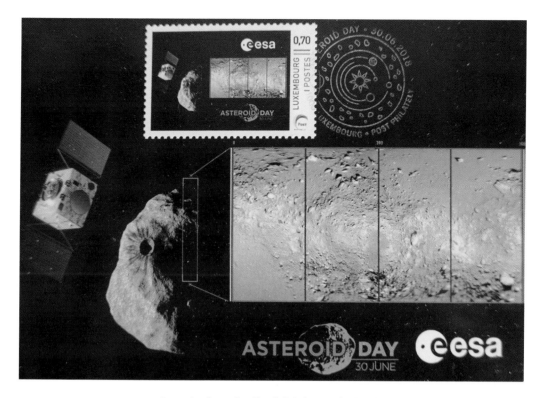

This stamp issued by Luxembourg in 2018 celebrates Asteroid Day, an annual international celebration of asteroid research research held on 30 June. The Asteroid Foundation is headquartered in Luxembourg.

I worked on the final, high resolution image. I created the basic asteroid in a digital 3D program, then did a 'render to disk' via that program to create a TIFF file. I then took the TIFF into Photoshop, where I fine tuned it and made it look more realistic. The star background originated from a traditionally painted star field that I scanned and used in Photoshop as the bottom (background) layer. Additional layers were added for the other elements in the art. Once Dr Marchis gave final approval I submitted the file for publication.

Finally, celebrate Asteroid Day every year on 30 June, the anniversary of the Tunguska event. First observed in 2014, it received sanction from the United Nations in 2016 and sponsors events worldwide.

WEBSITES

Websites to consult for further reading about asteroid studies are:

Asteroids and Remote Planets section of the British Astronomical Association:
www.britastro.org/section_front/8

A Guide to Minor Planet photometry:
www.minorplanet.info/ObsGuides/Misc/photometryguide.htm

A citizen science project associated with the Osiris Rex spacecraft mission to Bennu:
www.asteroidmission.org/get-involved/target-asteroids

The Planetary Society has a page dedicated to defending Earth against impacts:
www.planetary.org/explore/projects/planetary-defense

The non-profit B612 Foundation is also dedicated to the discovery and deflection of asteroids: https://b612foundation.org

Follow the work of the Spacewatch project:
https://spacewatch.lpl.arizona.edu

View imagery and perform analysis on orbiter data from Ceres:
https://trek.nasa.gov/ceres/index.html

Current information:
https://solarsystem.nasa.gov/asteroids-comets-and-meteors/
asteroids

The International Asteroid Warning Network: iawn.net

Learn about the development of war games to save Earth from
asteroid impacts on the website of the Planetary Defense
Conference: pdc.iaaweb.org

Homepage of the Minor Planet Center:
https://minorplanetcenter.net

The Jet Propulsion Laboratory (JPL) hosts the HORIZONS
system, which gives access to key solar system data and accurate
ephemerides for all known asteroids:
https://ssd.jpl.nasa.gov/?horizons

JPL also has a website for the Center for Near-Earth Object Studies
(CNEOS):
https://cneos.jpl.nasa.gov

For a European perspective, the European Space Agency's NEO
Coordination Centre publishes a monthly newsletter:
neo.ssa.esa.int

Recent discoveries can also be found on the Virtual Telescope
Project:
https://www.virtualtelescope.eu

The Meteoritical Society, in association with the Lunar and Planetary Institute, has a searchable database about all known meteorites: www.lpi.usra.edu/meteor

REFERENCES

1 THE STRANGER CERES

1 William Oxley, *Angelic Revelations Concerning the Origin, Ultimation, and Destiny of the Human Spirit*, vol. II (Manchester, 1877).

2 Philip T. Metzger et al., 'The Reclassification of Asteroids from Planets to Non-planets', *Icarus*, CCCXVIII (2018), pp. 21–32.

3 Clifford J. Cunningham, *Discovery of the First Asteroid, Ceres* (Cham, 2015), pp. 26–9.

4 Lorraine Daston, *Ravening Curiosity and Gawking Wonder in the Early Modern Study of Nature* (Berlin, 1994), p. 9.

5 Clifford J. Cunningham, *Studies of Pallas in the Early Nineteenth Century* (Cham, 2016).

6 Jean d'Alembert, 'Introduction au Traité de dynamique', *Oeuvres* I (1743), p. 392.

7 Martin Dyck, 'Mathematics and literature in the German Enlightenment', in *Studies on Voltaire and the Eighteenth Century*, CXC (Oxford, 1980), pp. 508–12.

8 Letter from Zach to Jan Sniadecki, 19 April 1802, Universytet Jagiellonik Biblioteka, Cracow.

9 John K. Rees, 'The Minor Planets', *School of Mines Quarterly*, XIX (1898), p. 251.

10 Luca Invernizzi, Alessandro Manara and Piero Sicoli, eds, *L'Astronomo Valtellinese Giuseppe Piazzi e la scoperta di Cerere* (Sondrio, 2001), p. 138.

11 Clifford J. Cunningham, *Early Studies of Ceres, and the Discovery of Pallas* (Cham, 2016).

12 James Dean, 'An Oration on Curiosity', *Port Folio*, VI/5 (New York, 1811), pp. 431–44.

13 Philo, *Philo*, trans. Francis Henry Colson, George Herbert Whitaker and Ralph Marcus (Cambridge, MA, 1929), vol. I, p. 57.

14 Clifford J. Cunningham, *Bode's Law and the Discovery of Juno* (Cham, 2017).

15 Sarah Fielding and Jane Collier, *The Cry: A New Dramatic Fable* (London, 1754).

16 Joseph Banks, letter to William Herschel, 2 June 1802, Royal Astronomical
   Society, B.42.
17 Franz Zach, 'Ueber die Pallas', *Monatlichte Correspondenz*, VI (1802), p. 90.
18 William Herschel, letter to Joseph Banks, 10 June 1802, Dawson Turner
   Collection XIII, 163–4, Natural History Museum, London.
19 Thomas Chrowder Chamberlin, *The Two Solar Families* (Chicago, IL, 1928), p. 243.
20 See *The Annual Cyclopedia* (New York, 1901), p. 44.
21 Wilhelm Olbers, 'Beobachtungen der Juno und Pallas', *Berliner Astronomisches
   Jahrbuch für das Jahr 1808* (1805), p. 179.
22 Wilhelm Olbers, *Berliner Astronomisches Jahrbuch*, p. 181.

2 WHERE ARE THE ASTEROIDS?

1 Aleksei N. Savich, *Historical View of the Discovery of Minor Planets or Asteroids* (in
   Russian) (St Petersburg, 1855).
2 This text is taken from unpublished material by Dr Tom Gehrels, given to me
   in the 1980s.
3 Victor Knorre, 'The Asteroids', *Pearson's Magazine*, X (1900).
4 William Fitzgerald, 'Sir Howard Grubb', *Strand Magazine*, XII (1896), p. 373.
5 Dan F. Seeson, 'The Minor Planets', *British Space Fiction Magazine*, II/4 (1955),
   pp. 30–35.
6 Guillermo Stenborg, Johnathan Stauffer and Russell Howard, 'Evidence
   for a Circumsolar Dust Ring Near Mercury's Orbit', *Astrophysical Journal*,
   DCCCLXVIII/I (2019).
7 Petr Pokorný and Marc Kuchner, 'Co-orbital Asteroids as the Source of Venus's
   Zodiacal Dust Ring', *Astrophysical Journal Letters*, DCCCLXXIII/2 (2019), L16.
8 Clifford J. Cunningham, '2060 Chiron Observations at Lowell Observatory',
   IAU Circular no. 4579, 13 April (1988).
9 Olivier R. Hainaut et al., 'Disintegration of Active Asteroid P/2016 G1',
   Astronomy and Astrophysics, DCXXVIII (2019).
10 Antranik Sefilian and Jihad R. Touma, 'Shepherding in a Self-gravitating Disk
   of Trans-Neptunian Objects', *Astronomical Journal*, CLVII/2 (2019).
11 Chris Lintott, 'To Ultima Thule and Beyond', www.lrb.co.uk, 2 January 2019.
12 See, for more on these studies and the researchers quoted in this paragraph,
   John R. Spencer et al., 'The Geology and Geophysics of Kuiper Belt Object
   (486958) Arrokoth', *Science*, CCCLXVII (2020).
13 Alan Stern et al., 'Initial Results from the New Horizons Exploration of 2014
   MU$_{69}$, a Small Kuiper Belt Object', *Science*, CCCLXIV/6441 (2019).
14 See Scott Sheppard quoted in George Dvorsky, 'Extreme Dwarf Planet
   FarFarOut Could Be the Most Distant Known Object in The Solar System',
   www.gizmodo.com.au, 1 March 2019.

15 Duncan H. Forgan, *Solving Fermi's Paradox* (Cambridge, 2019), pp. 338–9.

16 Shmuel Bialy and Abraham Loeb, ''Could Solar Radiation Pressure Explain 'Oumuamua's Peculiar Acceleration?', *Astrophysical Journal Letters*, DCCCLXVIII (2018).

17 Dimitri Veras and Daniel J. Scheeres, 'Post-main-sequence Debris from Rotation-induced YORP Break-up of Small Bodies', *Monthly Notices of the Royal Astronomical Society*, CDXCII/2 (2019), pp. 2437–45.

18 Clifford J. Cunningham, *Introduction to Asteroids* (Richmond, VA, 1988).

19 P. Vernazza et al., 'A Basin-free Spherical Shape as an Outcome of a Giant Impact on Asteroid Hygiea', *Nature Astronomy*, IV (2020), pp. 136–41.

20 Sota Arakawa, Ryuki Hyodo and Hidenori Genda, 'Early Formation of Moons around Large Trans-Neptunian Objects via Giant Impacts', *Nature Astronomy*, III (2019), pp. 802–7.

21 Tom Seccull et al., '2004 EW$_{95}$: A Phyllosilicate-bearing Carbonaceous Asteroid in the Kuiper Belt', *Astrophysical Journal Letters*, DCCCLV/2 (2018), p. L26.

22 Rogerio Deienno et al., 'Is the Grand Tack Model Compatible with the Orbital Distribution of Main Belt Asteroids?', *Icarus*, CCLXXII (2016), pp. 114–24.

23 Simona Pirani et al., 'Consequences of Planetary Migration on the Minor Bodies of the Early Solar System', *Astronomy and Astrophysics*, DCXXIII (2019), pp. A169.

24 Paul E. Olsen et al., 'Mapping Solar System Chaos with the Geological Orrery', *Proceedings of the National Academy of Sciences*, CXVI (2019), pp. 10664–73.

3 EXPLOSIONS AND IMPACTS

1 Isaac Facio, 'Fabric of the Universe: Exploring the Cosmic Web in 3-dimensional Woven Textiles', *Biennial History of Astronomy Workshop* (University of Notre Dame, 2019).

2 Clifford J. Cunningham, *Investigating the Origin of the Asteroids and Early Study of Vesta* (Cham, 2017).

3 *Macphail's Edinburgh Ecclesiastical Journal*, LI, Notes on Theology and Science, no. 8 (1850), pp. 173–83

4 Johann Schroeter, *Berliner Astronomisches Jahrbuch* (1807).

5 Nicholas Michell, *The Immortals: or, Glimpses of Paradise* (London, 1876).

6 Michaël Marsset et al., 'The Violent Collisional History of Aqueously Evolved (2) Pallas', *Nature Astronomy* (2020).

7 Kelsi N. Singer et al., 'Impact Craters on Pluto and Charon Indicate a Deficit of Small Kuiper Belt Objects', *Science*, CCCLXIII/6430 (2019), pp. 955–9.

8 William K. Hartmann, 'History of the Terminal Cataclysm Paradigm', *Geosciences*, IX/7 (2019).

9 Moser, D. E. et al., 'Decline of Giant Impacts on Mars by 4.48 Billion Years Ago and an Early Opportunity for Habitability', *Nature Geoscience*, XII (2019), pp. 522–7.

10 Robert DePalma et al., 'A Seismically Induced Onshore Surge Deposit at the Kpg Boundary, North Dakota', *Proceedings of the National Academy of Sciences*, CXVI (2019), pp. 8190–99.

11 Pincelli Hull et al., 'On Impact and Volcanism across the Cretaceous-Paleogene Boundary', *Science*, CCCLXVII/6475 (2020), pp. 266–72.

12 Lorien F. Wheeler et al., 'Probabilistic Assessment of Tunguska-scale Asteroid Impacts', *Icarus*, CCCXXVII (2019), pp. 83–96.

13 Mario Pino et al., 'Sedimentary Record from Patagonia, Southern Chile Supports Cosmic-impact Triggering of Biomass Burning, Climate Change and Megafaunal Extinctions at 12.8ka', *Nature Scientific Reports*, IX/I (2019).

14 Andrew M. T. Moore et al., 'Evidence of Cosmic Impact at Aby Hureyra, Syria at the Younger Dryas Onset (~12.8 ka): High-temperature melting at >2200 °C', *Scientific Reports*, X/I (2020).

15 Martin B. Sweatman and Dimitrios Tsikritsis, 'Decoding Göbekli Tepe with Archeoastronomy: What Does the Fox Say?', *Mediterranean Archaeology and Archaeometry*, XVII/I (2017), pp. 233–50.

16 J. Francis Thackeray, Louis Scott and P. Petrerse, 'The Younger Dryas Interval at Wonderkrater (South Africa) in the Context of a Platinum Anomaly', *Palaeontologia Africana*, LIV (2019), pp. 30–35.

17 P. J. Silvia et al., *The 3.7kaBP Middle Ghor event: Catastrophic Termination of a Bronze Age Civilization*, American Schools of Oriental Research annual meeting, Denver, 17 November 2018.

18 Sung Wook Paek et al., 'Optimization and Decision-making Framework for Multi-staged Asteroid Deflection Campaigns under Epistemic Uncertainties', *Acta Astronautica*, CLXVII (2020), pp. 23–41.

19 David Kramer, 'DOE Prepares Major Upgrade of its Lithium-6 Operations', *Physics Today*, LXXI (2018), pp. 29–31.

20 Toshihiro Kasuga et al., 'A Fireball and Potentially Hazardous Binary Near-Earth Asteroid (164121) 2003 YT$_1$', *Astronomical Journal*, CLIX/2 (2020).

21 Gerrit Budde, Christoph Burkhardt and Thorsten Kleine, 'Molybdenum Isotopic Evidence for the Late Accretion of Outer Solar System Material to Earth', *Nature Astronomy*, III (2019), pp. 736–41.

22 Bettina Schaefer et al., 'Microbial Life in the Nascent Chicxulub Crater', *Geology*, XLVIII (2020).

23 Rafael Navarro-Gonzalez et al., 'Abiotic Input of Fixed Nitrogen by Bolide Impacts to Gale Crater During the Hesperian', *Journal of Geophysical Research: Planets*, CXXIV/I (2019), pp. 94–113.

24 Eva Stüeken et al., 'Nitrogen Isotope Ratios Trace High-pH Conditions in a Terrestrial Mars Analog Site', *Science Advances*, VI/9 (2020).

25 Michael Nuevo, George Cooper and Scott Sandford, 'Deoxyribose and Deoxysugar Derivatives from Photoprocessed Astrophysical Ice Analogues and Comparison to Meteorites', *Nature Communications*, IX (2018), pp. 1–10.

26 Philipp R. Heck et al., 'Lifetimes of Interstellar Dust from Cosmic Ray Exposure Ages of Presolar Silicon Carbide', *Proceedings of the National Academy of Sciences*, CXVII/4 (2020), pp. 1884–9.

27 Kenneth Amor et al., 'The Mesoproterozoic Stac Fada Proximal Ejecta Blanket, NW Scotland', *Journal of the Geological Society*, CLXXVI (2019), pp. 830–46.

28 Sara Mazrouei et al., 'Earth and Moon Impact Flux Increased at the End of the Paleozoic', *Science*, CCCLXIII/6424 (2019), pp. 253–7.

29 Timmons M. Erickson et al., 'Precise Radiometric Age Establishes Yarrabubba, Western Australia, as Earth's Oldest Recognised Meteorite Impact Structure', *Nature Communications*, XI (2020).

30 Alexander von Humboldt, *Cosmos: A sketch of a physical description of the Universe*, trans E. C. Otté (London, 1849) vol. I, p. 97.

31 Nevil S. Maskelyne, 'Whence Come Meteorites?', *Nature London*, II (1870), pp. 77–8.

32 Richard C. Greenwood, Thomas H. Burbine and Ian A. French, 'Linking Asteroids and Meteorites to the Primordial Planetesimal Population', *Geochimica et Cosmochimica Acta*, CCLXXVII (2020), pp. 377–406.

33 Wilhelm Olbers, *Astronomische Nachrichten*, III (1824), pp. 5–10.

34 Larry R. Nittler et al., 'A Cometary Building Block in a Primitive Asteroidal Meteorite', *Nature Astronomy*, III (2019), pp. 659–66.

4 ASTEROIDS IN POPULAR CULTURE

1 'Present State of Commerce in Books, with remarks on the love of reading, in the interior of Russia', *Literary Panorama*, vol. I (London, 1807), pp. 145–8.

2 Enoch Fitzwhistler, 'A Modern Astronomer', *The Carpet-bag*, I/36 (1851), p. 2.

3 Clifford J. Cunningham, *Bode's Law and the Discovery of Juno* (Cham, 2017), p. 99.

4 John F. Bowers, 'James Moriarty: A Forgotten Mathematician', *New Scientist*, CXXIV/1696 (1989), pp. 17–19.

5 Charles Bucke, *On the Beauties, Harmonies and Sublimities of Nature*, vol. II (London, 1823).

5 Visiting an Asteroid

1 Marc Rayman, 'NASA Engineer Marc Rayman Taps Spock's Brain for Deep Space 1 Mission', www.startrek.com/videos, 2019.
2 Fred Jourdan et al., 'Timing of the Magmatic Activity and Upper Crustal Cooling of the Differentiated asteroid 4 Vesta', *Geochimica et Cosmochimica Acta*, CCDXXIII (2020), pp. 205–25.
3 Ziliang Jin and Maitrayee Bose, 'New Clues to Ancient Water on Itokawa', *Science Advances Acta Astronautica*, V/5 (2019).
4 Dante S. Lauretta et al., 'Episodes of Particle Ejection from the Surface of the Active Asteroid (101955) Bennu', *Science*, CCCLXVI/6470 (2019).
5 Carl W. Hergenrother et al., 'The Operational Environment and Rotational Acceleration of Asteroid (101955) Bennu from OSIRIS-REx Observations', *Nature Communications*, X/1 (2019).
6 Jacob Abrahams and Francis Nimmo, 'Ferrovolcanism: Iron Volcanism on Metallic Asteroids', *Geophysical Research Letters*, XLVI/10 (2019), pp. 5055–64.
7 Greg Wilburn et al., 'Guidance, Navigation and Control of Asteroid Mobile Imager and Geologic Observer (AMIGO)', ArXiv 1902.02071 (2019).
8 John S. Lewis, *Asteroid Mining 101: Wealth for the New Space Economy* (Moffett Field, CA, 2015).
9 Martin Elvis and Tony Milligan, 'How Much of the Solar System Should We Leave as Wilderness?', *Acta Astronautica*, CLXII (2019), pp. 574–80.

Appendix: Engaging with Asteroids

1 Francis Rolt-Wheeler, ed., *The Science-history of the Universe* (New York, 1909), p. 226.
2 Peter Broughton, *Northern Star: J.S. Plaskett* (Toronto, 2018).
3 'We Were Prisoners on Beast Asteroid', *House of Mystery*, no. 113 (1961), DC Comics.
4 Gary Morris, 'Empower Citizen Science to the Benefit of Us All', *Austin American-Statesman*, 12 April 2019, p. A13.
5 Huub de Groot, 'Occultations and Eclipses in Asteroid Systems', *Journal for Occultation Astronomy*, IX/2 (2019), pp. 3–10.

SELECT BIBLIOGRAPHY

Abreu, Neyda, ed., *Primitive Meteorites and Asteroids* (Amsterdam, 2018)

Alvarez, Walter, *T. Rex and the Crater of Doom* (Princeton, NJ, 1997)

Badescu, Viorel, ed., *Asteroids: Prospective Energy and Material Resources* (Heidelberg, 2013)

Barnes-Svarney, Patricia, *Asteroid: Earth Destroyer or New Frontier?* (New York, 1996)

Burbine, Thomas H., *Asteroids: Astronomical and Geological Bodies* (Cambridge, 2017)

Cox, Donald William, and James H. Chestek, *Doomsday Asteroid* (New York, 1998)

Davies, John Keith, *Cosmic Impact* (New York, 1986)

Elkins-Tanton, Linda, and Benjamin Weiss, eds, *Planetesimals* (Cambridge, 2017)

Frankel, Charles, *The End of the Dinosaurs: Chicxulub Crater and Mass Extinctions* (Cambridge, 1999)

Hamilton, Thomas William, *Dwarf Planets and Asteroids* (Houston, TX, 2014)

Lewis, John S., *Rain of Iron and Ice* (New York, 1996)

McCall, Gerald J. H., Alan J. Bowden and Richard J. Howarth, eds, *The History of Meteoritics and Key Meteorite Collections: Fireballs, Falls and Finds* (London, 2006)

Matsuoka, Ayako, and Christopher T. Russell, eds, *Hayabusa 2* (Cham, 2018)

Peebles, Curtis, *Asteroids: A History* (Washington, DC, 2000)

Schmidt, Nikola, ed., *Planetary Defense: Global Collaboration for Defending Earth from Asteroids and Comets* (Cham, 2018)

Starkey, Natalie, *Catching Stardust: Comets, Asteroids and the Birth of the Solar System* (London, 2018)

Steel, Daniel, *Target Earth: The Search for Rogue Asteroids* (New York, 2000)

Trigo-Rodriguez, Josep, Maria Gritsevich and Herbert Palme, eds, *Assessment and Mitigation of Asteroid Impact Hazards* (Cham, 2017)

Verschuur, Gerrit L., *Impact! The Threat of Comets and Asteroids* (Oxford, 1996)

Yeomans, Donald K., *Near-Earth Objects* (Princeton, NJ, 2013)

ACKNOWLEDGEMENTS

Thanks to Thomas Burbine (Mount Holyoke College) who invited me to address his students on asteroid history in 2018; Carlo Caruso (Sienna University) for his translation of the Piazzi poem; Lynette Cook for her artwork of Patroclus and its moon; Sally Poor for letting me photograph her late husband's artwork; Archer Chen of the Taipei Astronomical Museum, who kindly arranged permission for me to use the illustration of the planetesimals during my 2019 visit to Taipei; Jim Scotti for his astronaut painting and photographs of both Gehrels and the Bok telescope; Larry Denneau for the image of Gault; Sonia Fernandez at the University of California Santa Barbara for giving me the illustration by Jennifer Rice; Dale Cruikshank for the 1979 photo at the Infrared Telescope; Robert DePalma for the dinosaur-related illustrations; and Ellen Howell for her tour in 2020 of the OSIRIS-REX Science Operations Centre in Tucson, and the collection of meteorites housed there. I am grateful for Robert Marcialis for his review of my section on Hilda and Centaur asteroids; comments by Steven Weinberg; William Sheehan for facilitating my introduction to Reaktion Books, and my Reaktion Press editor Peter Morris for seeing this book through its development; Wayne Orchiston for overseeing my PhD thesis on asteroids; and a unique remembrance of MPC director Brian Marsden, my thesis advisor and a great inspiration. Thanks also for comments made by the following at the Near-Earth Asteroid workshop I attended in December 2019 at the University of Arizona in Tucson: Vishnu Reddy, Richard Wainscoat, Mario Juric, Amy Mainzer, Bryce Bolin and Carl Hergenrother. An important thanks to those who contributed their own unique reflections on asteroid science with passages written specifically for this book: Timmons Erickson, Robert McMillan, Gordon Osinski, Jordan Marche, Franck Marchis, Simona Pirani, Marc Rayman, Roy Tucker, and last but not least, Bill Hartmann and Don Davis, who also helped me with my first asteroid book in the 1980s. Finally, a personal thanks to Gareth Williams, associate director of the Minor Planet Center, for writing the Preface.

Photo Acknowledgements

The author and publishers wish to express their thanks to the below sources of illustrative material and/or permission to reproduce it.

Courtesy of Andrews McMeel Syndication: p. 63; Arizona State University: p. 150 bottom; author's collection: pp. 32, 33, 119, 120, 121, 167; Lynette Cook and W. M. Keck Observatory: p. 67; Clifford J. Cunningham: pp. 58, 91, 108; ESA: pp. 148 bottom (ScienceOffice.org), 169 (ScienceOffice.org/Luxembourg Post); ESO: pp. 48 top (L. Calçada and Nick Risinger (skysurvey.org)), 60 (L. Calçada and Nick Risinger (skysurvey.org)), 61 (M. Kornmesser), 68, 69 (M. Kommesser), 71 (J. L. Dauvergne & G. Hüdepohl (atacamaphoto.com), 82 (Swinburne Astronomy Productions), 83 (A. Mueller, MPIA), 102 (Gaia/DPAC), 104 (JAXA); Fermilab: p. 103 (Reidar Hahn); © William Hartmann, used with kind permission of the artist: p. 48 centre right; JAXA: p. 141 (Akihiro Ikeshita); JAXA, University of Tokyo, Kochi University, Rikkyo University, Nagoya University, Chiba Institute of Technology, Meiji University, University of Aizu, AIST: pp. 139, 140; LSST Project/ NSF/AURA: p. 101; Lunar and Planetary Institute/USRA: p. 109; NASA: pp. 39, 148 top, 152, 162; NASA, ESA, K. Meech and J. Kleyna (University of Hawaii), and O. Hainaut (European Southern Observatory): p. 50; NASA/Goddard/University of Arizona: pp. 143, 144; NASA/Goddard/University of Arizona/Lockheed Martin: p. 145; NASA/JHUAPL: p. 147; NASA/JHUAPL/SwRI: p. 55; NASA/JHUAPL/ Southwest Research Institute/Roman Tkachenko: p. 57; NASA/JPL-Caltech: pp. 41, 97; NASA/JPL-Caltech/Arizona State Univ./Space Systems Loral/Peter Rubin: p. 150 top; NASA/JPL-Caltech/UCLA: pp. 45 top, 98; NASA/JPL-Caltech/ UCLA/MPS/DLR/IDA: pp. 130, 131, 132, 133, 134, 137 top left and top right; NASA/ JPL-Caltech/UCLA/MPS/DLR/IDA/PSI: pp. 135, 136; NASA/JPL-Caltech/UCLA/MPS/ ISA/ASI/INAF: p. 137 centre left; NASA/JPL-Caltech/University of Arizona: p. 106; NASA/Denise Watt: p. 154; Photo courtesy of Dr Gordon Osinski: p. 165; Courtesy Planetary Society: p. 129 (Data from NASA/JPL/JHUAPL/UMD/JAXA/ESA/OSIRIS

INDEX

Page numbers in **bold italics** refer to illustrations